T0314252

THE COMMONS IN AN AGE OF UNCERTAINTY

The Commons in an Age of Uncertainty

Decolonizing Nature, Economy, and Society

FRANKLIN OBENG-ODOOM

UNIVERSITY OF TORONTO PRESS
Toronto Buffalo London

ISBN 978-1-4875-0176-1 (cloth)
ISBN 978-1-4875-1390-0 (PDF)
ISBN 978-1-4875-3761-6 (EPUB)

Library and Archives Canada Cataloguing in Publication

Title: The commons in an age of uncertainty : decolonizing nature,
economy, and society / Franklin Obeng-Odoom.
Names: Obeng-Odoom, Franklin, author.
Description: Includes bibliographical references and index.
Identifiers: Canadiana (print) 20200212427 | Canadiana (ebook)
2020021246X | ISBN 9781487501761 (hardcover) |
ISBN 9781487513900 (PDF) | ISBN 9781487537616 (EPUB)
Subjects: LCSH: Commons. | LCSH: Public lands. | LCSH: Right
of property. | LCSH: Decolonization.
Classification: LCC HD1286 .O24 2020 | DDC 333.2 – dc23

University of Toronto Press acknowledges the financial assistance to its
publishing program of the Canada Council for the Arts and the Ontario Arts
Council, an agency of the Government of Ontario.

 Canada Council   Conseil des Arts
for the Arts   du Canada

 ONTARIO ARTS COUNCIL
CONSEIL DES ARTS DE L'ONTARIO
an Ontario government agency
un organisme du gouvernement de l'Ontario

Funded by the   Financé par le
Government   gouvernement
of Canada   du Canada
 Canada

Dedicated to the Original Land Economists whose vision, if revisited and revamped, would generate the much-needed certainty for our age, a possible alternative, on which visionaries like Anne Haila (1953–2019) worked until she returned to the land.

Contents

Preface

This book develops an uncomfortable treatise. The social problems we face in our age of uncertainty have been misdiagnosed. What to do about them is misinformed. These contentions question both positivism and progressivism. Enclosure is a tragedy, but the left-wing proposition that "commoning" everything would solve global problems, as Peter Linebaugh (2008), a leading scholar on the left, contends, is problematic.

The treatise is shaped by an unwavering commitment to original land economics. Although now gagged by vested interests, its liberation is needed to understand the political economic realities of our times, to demonstrate the poverty of existing analyses, and to shape the road to a new world order. The book treats land as the problem as well as the solution, systematically demonstrating why this is and how it became this way. It is a testimony to why we need a world where land is common and the ways in which it can be achieved.

Given the radical principles on which this book is based, it is worth telling a brief story about the origins in my own life. It is not just that "the personal is political economic," as feminists have taught us, but that the political economic is also personal. A combination of a land monopoly and related famine, political dictatorship, and general economic depression in the 1970s and 1980s forced many people out of Ghana, including my parents, to look for work.

I had to grow up with my grandfather, a post-colonial judge doubling as a choirmaster and an organist in the Methodist Church. He would take me to work, to his court and to the church. Delivering justice in the English adversarial court system in a West African environment of anger and hunger along with teaching hymns of hope and humanity must have been difficult. In his autobiography, he wrote that "the wonder of the places I worked was how a judge could also be a choir-master, and an organist all at a go" (Abakah, 2006, p. 7). My grandfather would stay up late to write his judgments under the prying eyes of a police officer provided by the state officially to protect him, but I wondered later whether the officer was there simultaneously to monitor him. Regardless, he wrote judgments that the Judicial Service of Ghana (2006) described as "brilliant, just and fair" (p. 18).

In a period called "The Lost Decade" in development economics, informal economic activities such as urban agriculture and urban street trading flourished. I, too, had to contribute to the collective income of the family by selling in the market as a school boy, helping to work on mixed cropped farms we made in the city, tending livestock, and occasionally helping on external family cocoa farms in the country to shore up our family income. An unintended consequence was that my agricultural science classes became more meaningful. I also had to share an increasingly crowded house in the city with family members from the country who would visit and work in informal economies to support their own mostly rural incomes.

These experiences of rural-urban interlinkages; interdependencies between formal and informal economies; the intersectionality of individual, public, and collective space; and justice in and out of court tampered with mercy and hope in the church helped me to develop three interlinked core values. I believe in both diversity and pluralism. I have an unwavering commitment to the centrality of land to our well-being, and I embrace justice as tangible, not as moralizing or empty talk.

Declining the offer of admission to study for a bachelor of arts in (mainstream) economics at the University of Ghana, I opted instead for a bachelor of science in land economy, which as I understood it, combined law and economics to deconstruct land: a fundamental backbone of the economy. While studying land economy, I was introduced to the work of thinkers such as Raleigh Barlowe, the urban land economist whose ideas would contribute to inspiring me to study urban economics.

A British Commonwealth Scholarship enabled me to realize this aspiration because it would fund my education in England. Studying urban economics in the transdisciplinary Development Planning Unit of the Bartlett School of Built Environment broadened my view of urban economics, my only regret being that it was too brief. The course ended at the master's level. I had no option for a PhD, which was what I wanted to do.

Finding a PhD supervisor was extremely difficult. The economics professors were neither interested in cities nor keen about putting justice at the centre of their work. They were even less inclined to consider land as central to economic analysis, perhaps because they thought it was substitutable with capital, as the mainstream economist and Nobel Laureate Theodore Schultz (1951) had strongly argued. At a deeper level, the more transdisciplinary an economist, the less respect was obtained. For example, a serious-minded economist, a friend, who became very interested in feminist analysis and wanted to embrace it, was advised against doing so for the reason that other economists would look down on his work. Similar accounts exist about specializing in the political economy of race.

Luckily, though, a few economists resist this professional inbreeding. I found one such person in the leading Australian political economist Frank Stilwell, who doubled as the father of insurrection in the University of Sydney economics department. A revolution against pedagogical monism, the mutiny resulted in the institution of intellectual holism in a new pluralist economics department called Department of Political Economy (Butler et al., 2009). It stood for what I wanted. With the invitation by Professor Stilwell to do a PhD under his supervision and a full scholarship from the university, I went off to Australia to study with Frank. I completed the study, and was recruited to research and teach property economics at the University of Technology Sydney.

There, I had the opportunity not only to learn, but also to teach, land economics. Alas! I was faced with an uphill task. The tradition of critical land economics that had drawn me to the field was bullet-riddled with complaints by students, low student ratings, disaffection by colleagues, and regular threats by the leadership of the university to discontinue it, banish it from the program, and forever bury it. I had to re-study and redesign the subject.

As a first step, I spent time at the Henry George School of Social Sciences in Chicago, which was intellectually rewarding and socially gratifying, as were my studies and socialization with land economists from the Association for Good Government. Within the academy, I had lengthy discussions with other political economists, including Frank Stilwell, Garrick Small, and John Pullen (Obeng-Odoom, 2017a). I developed a wider appreciation of the range of theoretical positions in land economics. The people with whom I consulted differed in their approaches, but they were all critical of mainstream land economists. I learned from their investigations into the commoning of land in our cities, peri-urban spaces, and rural areas. Together with probing complex property rights, which I studied with Spike Boydell, Australia's leading property theorist, then the director of the Asia-Pacific Centre for Complex Real Property Rights, I worked on the outlines of the analyses of the commons. Inspired by these influences, I worked on new ways of deconstructing the status quo, more effectively framing and more comprehensively studying the nature of land in non-European settings of Africa. In such societies, land is not only different but also socially differentiated from landed property rights systems forcibly commodified and imposed by colonizers.

Developing these insights in a coherent, systematic, and comprehensive study course was challenging. Yet eventually, significant collaboration with students led me to develop a commons-based curriculum. The results convincingly showed that a critical tradition of land economics is not only possible to develop, but that it is relevant, and highly appreciated by students and non-students alike. I published this breakthrough as innovative pedagogy in *International Journal of Pluralism and Economics Education* (Obeng-Odoom, 2017a), *International Review of Economics Education* (Obeng-Odoom, 2019a), and *Australian Universities' Review* (Obeng-Odoom, 2019b).

Building original research in this tradition, however, proved daunting. Political economists are quick to point out how difficult it is to publish their critical work in leading journals because of bias. They are less aware of the additional bias within and among their own against land economics. The dominant political economy today is centred on labour and capital, giving little or no attention to land. Indeed, even when political economists are concerned about "nature," their interest is on what they call "the second contradiction of capital" (O'Connor, 1988, 1991). Framing land economics as political economy, therefore, comes at a serious cost: it might well be successful in the classroom, but not on the pages of the leading political economy journals. This barrier is all the more striking when journals retain the label *land economics*, but not the content of original land economics (Obeng-Odoom & Bromley, 2020), while policy-oriented land journals look askance at this body of work.

In the end, a handful of committed editors, against the risk of burying their journals and magazines under the land, urged me on. Not only did they tolerate my attempt at revival and reconstruction, they encouraged it. Thus, my earlier analyses of the commons have appeared in journals such as the *American Journal of Economics and Sociology*, the leading radical land economics journal, whose editor, Clifford Cobb, personally encouraged me to develop my analyses of the commons. *Progress* published more popular versions of my writing, while I gained funding and the privilege of face-to-face advice from the editors associated with *Good Government: A Journal of Political, Social and Economic Comment.*

This book is the culmination of my struggles as a political economic analyst of the global system specializing in land economics. The vision of the book is to help re-engage and revamp original land economics, centrally focused on the intersectionality of socio-spatial, economic, environmental, and ecological justice. To realize this vision, the book has three key emphases. First, it tries to prioritize a particular type of political economy that develops pluralism not in the narrow sense of multidisciplinarity but in terms of the transdisciplinarity of ideas, diversity of voices, and a strong interest in policy and action. Second, its unit of analysis is not only the firm or the individual, but also the intersectionality of class, race, gender, and other institutions. Third, as a black economist with great respect for the land, including recognizing my own spiritual connection to it, I have written the book in such a way that it defends, holds, and keeps land as a major pillar for commons research.

Without the friendship, collegiality, and solidarity of several committed scholars, activists, and administrators who share this vision, this much needed, but widely resented, "minority report" would have continued to be censured. I must single out Clifford Cobb for thanks. Cliff gave me extensive comments and encouragement to develop my analysis of the commons, some of which he also generously published along the way. Not only has his support been unwavering, but it has also grown with every step I have taken to extend my analysis of the commons.

The referees for *Review of Social Economy*, *International Critical Thought*, and *Review of Radical Political Economics* deserve mention. Their feedback helped to strengthen the papers whose extended and revised versions serve as the foundation of this book. For several rounds of detailed constructive criticisms and suggestions for the entire manuscript, I would like to thank the reviewers for the University of Toronto Press. Thanks also to my editor, Jennifer DiDomenico, for facilitating the process and for being the advocate of the book in the internal circles and committees of the Press. I would like to say thank you to Dawn Hunter for important support and help. Many thanks to Nancy Wills for her feedback and for preparing the index. Thanks also to Leah Connor for helping to finalize the manuscript and processing the proofs. To the Austrian ecologist Andreas Exner of the Vienna University of Economics and Business Administration; to the Australian anthropologist, Robbie Peters of the University of Sydney; to Johannes Euler of the University of Duisburg-Essen, Germany; to Kim Shanna Neverson of the Aboriginal Health Service Organization, Montreal, Canada; and to Stéphane Nahrath of the University of Lausanne, Switzerland, I say thank you for your detailed constructive feedback, which you offered promptly and in the spirit of helping to advance the global research on the commons.

Being on a theme that engages, but also transcends, academic research, I obtained many insights from meetings with activists. In particular, the meeting organized by the Commons Strategies Group and the Heinrich Böll Foundation in Lehnin, Germany exposed me to various dimensions of the debate. Deep Dive as they called the meeting, brought me into contact with leading "commoners," such as Silke Helfrich, David Bollier, and Michel Bauwens, from whom I learned a great deal and for which I feel most grateful. I must emphasize and gratefully acknowledge my intellectual debts to the Swiss historian Daniel Schläppi (University of Bern in Switzerland) and British state theorist Bob Jessop (University of Lancaster) both of whom generously offered me guidance and advice. Dan, especially, maintained contact and offered me encouragement and feedback on aspects of the book for which I needed his razor-sharp mind.

In the lead up to actually writing the book, I received additional abundant support for which I express my appreciation. In particular, I would like to thank Bronwyn Clark-Colee, then research manager, Faculty of Design, Architecture and Building, at the University of Technology Sydney (UTS) in Australia for helpful suggestions and encouragement. Spike Boydell, formerly of the School of Built Environment, UTS, and Frank Stilwell of the Department of Political Economy at the University of Sydney inspired me to unite land and political economy, greatly facilitating my work on property and political economy, often at personal and professional cost to them. The Academy of Finland funded the Urban Land Tenure Project, pioneered by Anne Haila who, accordingly, became academy professor. As part of this project, the book has benefited from an Academy of Finland grant for which I am particularly grateful. Both the Henry George Foundation and the Association for Good Government in Australia offered generous funding for this book for which I am very grateful. I would like to thank Richard Giles and Faye Giles for their personal commitment to the successful completion of this book.

Annie Hero of the University of New South Wales deserves special mention and thanks for her unwavering support and abundant encouragement freely given even in the face of grave adversity. I appreciate her friendship and solidarity, as well as her intellectual counsel on crucial aspects of the book, especially chapter 1, where her many carefully considered questions, suggestions, and criticisms enabled me to simplify and more systematically present complex arguments. Thanks also to Mi Shih (Rutgers University), discussant of my paper at the 2019 AAG Conference, for her helpful comments, and to the late Anne Haila (University of Helsinki) for her leadership and inspiration in putting the panel on Alternatives to Private Landownership together. Her colleagues and students at the Helsinki School of Critical Urban Studies deserve thanks, too. As I have developed my analyses in the context of the critical development studies that I teach at the University of Helsinki, I must thank my colleagues in Development Studies and the Helsinki Institute of Sustainability Science for their collegiality. Kofi Baah–Kofi Boye deserves exceptional thanks for the personal stability and meaning he has brought into my own life of scarcely known, but ever-present, turmoil and uncertainty.

I hope this book provides further inspiration to develop the foundations of another world, stratification economics, and the outlines of a new ecological political economy centred on land.

Franklin Obeng-Odoom
University of Helsinki, Finland

PART A

The Problem

Chapter One

The Age of Uncertainty

Introduction

We live in "an Age of Uncertainty" (Galbraith, 1977). Such uncertainty arises today from many pressing social problems. Consider the COVID-19 pandemic. It has brought the world's most powerful nations to their knees. Both the learned and the lay have fallen ill and died. Celebrated cities are exposed, while celebrities have recoiled. Lockdowns have paved the way for crackdowns. Both Thorstein Veblen's leisure class and Karl Marx's working class have similarly been quarantined. The certainty of the past has dissolved into thin air. The threat of a warming planet; growing inequalities; the crises of migration; the rise of extreme leaders; the escalating tensions about the continued drilling and use of oil amid rising emission levels; and conflictual governance of the Strait of Hormuz and the South China Sea are other features of today's age of uncertainty. The intensifying power of technology and how it is leveraged by elites to increase their privilege add to the picture, along with the intriguing skill of the powerful to recast themselves as meek and weak while still controlling society. The concentration of more and more people in cities; the pressure on drinking water, indeed on water bodies more generally amid rising world production; ad-hoc political-economic uprisings around the world; and the resurgence of organized challenge to the establishment create dangers, possibilities, and uncertainties.

Since 1914, which, according to John Kenneth Galbraith (1977), marked the end of the age of certainty, the Age of Uncertainty has remained and indeed has worsened. The character of "the beginning of the Age of Uncertainty," Galbraith wrote, "derived ultimately from the new social alignments, the new governing coalition that now emerged" (Galbraith, 1977, p. 160). In Galbraith's time, those social alignments included how capitalist and propertied interests, along with their think tanks, media houses, and education institutions, put their resources together to advance their course. Uniting behind Milton Friedman, these forces used the power of wealth and technology to extensively transform the Global North and its beliefs, political, and economic systems (Burgin, 2013; Galbraith, 1977). Today, a major source of uncertainty is the political economy of the great transformation underway in the Global South. Galbraith (1977) discussed the colonial question and the problems of development and underdevelopment in *The Age of Uncertainty*, of course, but he framed these issues as reflective of the colonizers' social constructions and considered them as "digressions" (see Galbraith, 1977, pp. 111, 132). So, the "problems" of the colonies were socially framed to aid in the advancement of the metropolitan societies of the Global North, but they were not at all central to the grammar of uncertainty.

Yet the great transformation in the Global South is a socioecological construction, not just a social construction with ecological consequences (Ross, 2017, p. 4). The "problems" of the Global South have been framed not just in social terms but also in ecological ways that ensure it continues to be dominated by the Global North. This dialectical relationship has *deepened*, not *weakened*. Even more striking, the transformation is *central* not *marginal* to the Age of Uncertainty. Formerly associated with pristine nature, the Global South has become the home of extinction, a region of increasing inequality, mass pollution, and biodiversity loss (see, for example, Dawson, 2016; Moser, 2020; Obeng-Odoom, 2020). Its share of the earth has also increasingly become the private property of a few classes, races, transnational corporations (TNCs), and powerful nations in ways that raise questions about sovereignty (Cobb, 2016). These uncertainties are clearly not separate: they are intertwined and interlinked with dynamic processes and uncertainties in the Global North.

Beyond describing what is happening, how can we explain its causes? Analysts (see, for example, Dragun, 2001; Giles, 2015, 2016a, 2016b; Harvey, 2011; Ostrom, 1990) point to Garrett Hardin's well-known body of work, including "the case against helping the poor" (Hardin, 1974) and "the tragedy of the commons" (Hardin, 1968), as *exemplifying* the reasons for this great transformation and its uncertainties. Hardin describes the "horror of the commons" (1968, p. 1247) as being the total destruction not only of nature but also of humanity. The commons – that is, the frontier, anything or any process that is commonly owned or managed – according to Hardin, destroys the basis on which humans depend, through either over-exploitation or dumping of toxins. These problems, he argued, are exacerbated by population growth.

This diagnosis is all the more striking because the state cannot do anything about it. As Hardin put it, the state is the site of "arbitrary decisions of distant and irresponsible bureaucrats" (1968, p. 1247). So, the widely held alternative to the commons, the use of the state as a solution, is worse.

In his words, "the alternative to the commons is too horrifying to contemplate" (1968, p. 1247). The solution, according to Hardin, is to institutionalize private property in nature (1968, p. 1247). This market approach, he contended, is widely accepted by all, what he described as "mutually agreed" (1968, p. 1247). He recommended its widespread adoption wherever a commons situation exists, such as the governance of parks and gardens, nature reserves, and oceans. Private property, in essence, would not only rid society of its many problems and uncertainties but would also make society more prosperous, stable, and sustainable.

Private property, then, is the path to progress. A key concept used by proponents of Hardin's causal theory to unlock the gates of this Eldorado is transaction costs (Dagdeviren & Robertson, 2016; Klaes, 2000). This idea – allegedly developed by Ronald Coase (1960) in the *Journal of Law and Economics* and later summarized as the Coase theorem – has a particular meaning quite distinct from its more popular usage. According to proponents, while the costs of state regulation of private management are high, the assignment of private property rights reduces transaction costs to zero because it costs little or nothing for individuals with private property rights to work by themselves. Indeed, privatizing the commons guarantees that individuals will put their private property rights to the highest and best use (Zhang, 2018). "All private owners," Alchian and Demsetz (1973, p. 22) famously noted, "have strong incentives to use their property rights in the most valuable way." This "property right paradigm" (Alchian & Demsetz, 1973) also holds that, for these very reasons, over time, individuals naturally choose more private property over state and common property. In turn, the aggregation of such individual rational choices *naturally* leads to the privatization of the commons in society, which then enables the process of economic development through the "wise use" of nature (for a review, see de Soto, 2000; Jacobs, 1995).

The notion of "transaction costs" is also embedded in an epistemology of history that sees the privatization of nature as a natural path for all societies. Highly influential (see, for example, World Bank, 2003, 2016), this view frames the question about privatizing nature as when it will happen, not how or even whether it will, or should, happen at all. Much like the dominant existing explanations for the transition from socialism to capitalism (see, for example, Friedman, 2007; Fukuyama, 1989, 1992; Huntington, 1993, 1996), this framework has come to define how economists approach the history of the Global South.

This compression history, as it is widely called, is highly problematic. Characterized as prioritizing statistical over comprehensive historical analysis, compression history focuses on individual end points, not on the dynamic relationships that structure the endpoints. Indeed, the end point data – often taken on face value – are usually old, manipulated, and underpinned by highly suspect assumptions (Hillbom & Bolt, 2018). It is this compression history approach that led Dutch economist, Robbert Maseland, to declare that "colonialism is history" (Maseland, 2018), meaning that history no longer matters in analysing the political economy of the uncertainties in the Global South. More systematic and experienced historians (see, for example, Diop, 1967; Rodney, 1972/2011; Zouache, 2017a) have, however, shown the fallacies of such crude historiography. In turn, there has been a revival of calls for "decolonizing methodologies" (Smith, 2012).

Elinor Ostrom (1990) and those committed to her work (e.g., Frischmann, 2013; Pennington, 2012; Tarko, 2012, 2017), sometimes called the Bloomington School of New Institutionalism (see Aligica & Tarko, 2012, p. 237), claim to offer a different interpretation and pathway. They do so by showing that individuals can cooperate under certain rules, rewards, and punishments without any external authority (Ostrom et al., 1992). Commentaries and endorsements of Ostrom's approach have come from across diverse methodological and political positions (see, for example, Amadae, 2004, 2015; Gunn, 2015; Haller et al., 2019; Milonakis & Meramveliotakis, 2013).

In essence, however, the work of Hardin and Ostrom are similar in many respects (Obeng-Odoom, 2015a, 2015b, 2016d). Their *explanation* of socioecological crises is methodologically "commonist," or methodologically nationalist at best (e.g., Cobb, 2016; Cousins, 2007; Dagdeviren & Robertson, 2016; Dawson, 2016; Exner, 2014, 2015; Hodgson, 2014; Kepe, 2008; Metcalfe & Kepe, 2008; Okoth-Ogendo, 2003; Shipton, 2007, 2009, 2010; Sjaastad & Cousins, 2008; Yifeng, 2008; Zhang, 2018). They attribute socioecological crises to what pertains *within* the commons (Bromley, 2008; Haller et al., 2019; Obeng-Odoom & Bromley, 2020) or within the common pool, not across common property regimes (Obeng-Odoom & Bromley, 2020), or within the nation-state (for a discussion of methodological nationalism, see Gore, 1996).

These separatist analyses are problematic because they neglect enduring and increasing interactions across scales and institutions within the global system. Also, they share a pre-analytic suspicion of the state as a contributor to, and inherently incapable of, addressing socioecological crises, even when there is evidence that the state, under certain circumstances, can be an agent of positive change. Not only do both demure from theories of social justice, but also both adopt a compression approach to history. In turn, both tend to be cited as inspiration for advocating the privatization of nature, as can be seen in citations in the work of the development agency the World Bank (World Bank, 2016, p. 27).

Unsettled Research Questions: Towards a New Research Approach

In light of these problems, three questions need to be answered. First, what alternative explanations can be provided for the socioecological crises in the Global South? Second, in privatizing nature, what are the wider socioecological consequences for different groups in the Global South? Third, could the commons in the Global South facilitate, instead of hinder, wider multi-scalar inclusive socioecological progress and prosperity, as famously suggested in a number of classic studies? (See, for example, Asante, 1975; Ciriacy-Wantrup & Bishop, 1975; Haller et al., 2019.)

These questions are central to how resources are distributed and controlled. How they are answered and the resulting policy choices made based on their answers shape how various socioecological groups maintain their livelihoods while still contributing to the resolution of planetary problems. Indeed, they provide the opportunity to examine in what ways, as collective polities, the countries of the Global South maintain their sovereignty. The unsettled questions sit at the very heart of the political-economic idea of the "Global South." In turn, they have generated widespread research, both historically (see chapter 2) and in contemporary times (see, for example, Amin & Howell, 2016; Castree, 2008a, 2008b; Collier, 2008, 2009, 2010; Ostrom, 1990). However, many of these studies are unsatisfactory (Bromley, 2008; Hiedanpää & Bromley, 2016; Obeng-Odoom & Bromley, 2020). They do not systematically theorize the commons in terms of land rights. If they do, they focus mostly on internal not external political-economic institutions. Hence, they do not investigate the relationship between the internal and the external conditions of the commons.

Some analysts have tried to bridge this gap. Yet they do not give any special place to nature. Consider Johannes Euler's (2018) work. Echoing the approach of the Commons Strategies Group, a widely recognized collection of scholars and activists who work on the commons, he argues that the idea of the commons itself has to be reconceptualized in terms of a cluster of relations of production and sharing that operate on democratic principles, both politically and economically, and, hence, must repudiate core features of capitalism such as racism and patriarchy (Figure 1.1).

This reconceptualization is widely *adopted* (accepted as is) or *adapted* (accepted with only minor variations), as many studies (see, for example, Hughes, 2016; Papadimitropoulos, 2018; Zhang, 2018) show. Figure 1.1 emphasizes that commons are better defined jointly by their collective rejection of capitalism as we know it and, even more broadly,

Figure 1.1. The Consensus Approach to the Commons.

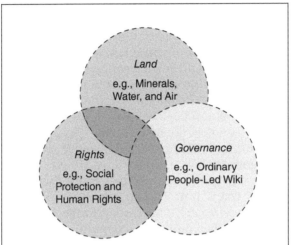

by their emphatic rejection of patriarchy and racism. Zhang (2018) points to the need for the commons to realize that humans, in fact, are not *homo sapiens* but rather *homo reciprocans*.

In other words, humans are more likely to be characterized by self-love, a desire to develop oneself without necessarily putting others down, not selfishness, a desire to advance by denigrating others or destroying the environment. These attributes are, of course, socially structured in ways that do not deny agency but could not be reduced to agency alone (Gronow, 1997) or selfishness alone, as is commonly claimed in mainstream economics. Some of these distinctions were made by Adam Smith himself (see, for example, Smith, 1776/2007) and, in more recent times by economists such as Daniel Kahneman (see, for example, Kahneman, 2011), himself hardly a radical political economist. However, they have been overlooked. More typically, the distinctions have been treated simplistically by modern mainstream economists and new institutional economists. Elinor Ostrom's approach is a case in point. Her framing of the commons as "common-pool resources" (Ostrom, 1990) has become the best known and the most influential in the social sciences, and yet it is also the most problematic partly because it overlooks the political economy of land, it is neglectful of justice, and it pays only scant attention to the social production of ecological crises (see chapter 3).

Can we not simply combine everything we know into one grand narrative, as suggested by Figure 1.1? According to Zhang (2018), all three elements of the concept of the commons must be considered together as in Figure 1.1. After all, "when analysing a commons of river we need to analyse it as an integrated commons that is composed of natural commons of water system, ecological diversity, environmental carrying capacity; institutional commons of welfare system in the river valley; and the cultural commons of identity, ethnicity and religious belief of the communities living in the river valley" (Zhang, 2018, p. 27).

In essence, then, there are two broad ontological approaches to seeking to address the "unsettled questions" posed earlier. The first is what, borrowing from J. K. Galbraith (1958/1998, pp. 6–17), I call the "Conventional Wisdom." It comprises the *essence of the analytical traditions* in which Hardin and Ostrom and their followers' work. The second broad approach is what I call the "Western Left Consensus." This approach is defined by the traditions that contend that everything that is collectivized and is anti-capitalist *is* or *ought to be* regarded as a commons and a solution to neoliberalism, which is used alongside the tragedy of the commons as key *explanatory frameworks* for explaining socioecological crises in the Global South (see, for example, Castree, 2008a, 2008b; Dunn, 2017; Fahnbulleh, 2020).

Nature in these existing explanations is not special but rather is one of many relations that can be commoned. When common, the aspiration is to develop socialism. Similarly, the privatization of nature must be considered analogous to the privatization of the products of state enterprises, for example. This social contract leads to an interesting, but quite limited, body of studies called the "neoliberalization of nature" (Castree, 2008a, 2008b). By these existing approaches, vague notions of the commons are invoked to refer to anything from capital and labour to land. All of them are given equal weighting. All of them are valid. All of them will build the silk road to socialism. Everything goes, perhaps, if certain human-designed rules are obeyed, and everything is threatened if those rules are violated by bad individual behaviour or by the malpractices of the state (see, for example, "The Climate Issue," 2019; Ostrom, 1990). The economy, especially the Indigenous economy, is also poorly theorized. Any activities that appear to use nature to satisfy material needs tend to be regarded as *destructive*. Religion is secularized. Spirituality is commodified. Both are vilified (Nelson, 2004, 2019). This framing of economy as outside nature or nature as outside economy continues to be one of the many "environmental heresies" in the Western Left Consensus (Hiedanpää & Bromley, 2016).

These approaches lead to profound confusion, and analyses based on them pass off symptoms as causes and causes as symptoms (see chapter 3). Analytically, they tend to conflate land and capital, treat land in

Eurocentric terms (e.g., its tendency to become a commodity), or mistake common property regimes as common pool resources (Obeng-Odoom & Bromley, 2020). As even advocates of such eclectic approaches admit (see, for example, Papadimitropoulos, 2018), the commons and capitalism have become so mutually dependent that, although developing the commons is admirable, defending the commons could become synonymous with defending capitalism itself because the commons can create the social basis for the further advancement of capitalism. Also, this existing literature typically neglects systematic conceptualization based on property rights and overlooks the long history of the commons, leading some of its advocates to claim that "commons are still in their infancy" (Papadimitropoulos, 2018, p. 317). Both the Conventional Wisdom and the Western Left Consensus are, therefore, unsatisfactory.

The Approach of This Book

This book develops and defends a third way, a Radical Alternative. It decolonizes the existing approaches (Smith, 2012). Decolonization, in this respect, means that I try to unravel the general and particular processes that have shaped and continue to shape social conditions in Africa. With the privatization of nature particularly sustained on the continent, the context for much of the contests on nature in the twenty-first century (see, for example Bromley, 2008; Okoth-Ogendo, 2003) must be made explicit.

I offer landed property relations as a research approach (Ely, 1917) that addresses the shortfalls of the Conventional Wisdom and the Western Left Consensus. My emphasis on land is important not just because it is one of the key factors of production but also because it is the very identity of Africans and many black societies around the world. Asserting this claim is easy; conceptualizing land is not. Indeed, many anthropologists claim that there is no such thing as a general concept of land in Africa because every neighbourhood has its own ideas of land. This view is, however, extreme. By studying the writings of the leading African and Africanist authorities on land (e.g., Asante, 1975; Cousins, 2007; Hill, 1961, 1966; Kepe, 2008; Maathai, 2004, 2011; Metcalfe & Kepe, 2008; Okoth-Ogendo, 2003; Shipton, 2007, 2009, 2010; Sjaastad & Cousins, 2008; Tonah, 2005), it is possible to conceptualize land as follows.

First, land is nature itself and nature is indistinguishable from economy, so trees and animals are considered in the context of land (especially as animals/trees are also commonly held among many Indigenous and African peoples). Land is the earth, water, and oil; indeed, all minerals whether "natural" or socially enhanced can be considered as land. Second, land protects, feeds, holds, and heals. So, land is sacred; it is to be revered and protected, indeed it is to be shared both intra- and inter-generationally. Third, there has always been a distinction between possession and ownership in land economics. In the Age of Uncertainty, the lines between the two have been blurred, in part because of the expansion of permanent buildings in the process of urbanization. Ownership, however, is quite distinct from possession even if they *appear* to be the same. Treating the two similarly can confuse freedom for bondage. One way to resolve this problem could be to rely on redistributive institutions, such as *abunu* (dividing a harvest into two) and *abusa* (dividing a harvest into three), which are concepts often applicable in land tenancies in the production process.

These ideas may sound vague, but they become concrete when juxtaposed with other concepts of land. For example, land is not separate from water/oil (compare with more Westernized notions of land in Li, 2014). Also, land is not labour or the products of labour, such as housing (compare with Western juridical pronouncements of land). Indeed, no amount of labour produces or justifies the appropriation of land (compare with the Lockean labour theory of land). Consequently, African Indigenous practices such as *abunu* and *abusa* reward labour separately and recognize that land conditions production. Land is not capital (compare with the neoclassical concept of substitutability, which considers land as capital).

This Africanist conception of land, therefore, clarifies the often-misleading view in mainstream economics that land is a commodity. If at all, land is simply made to appear as a commodity when, in fact, it is only a "fictitious commodity" (see Polanyi, 1944/2001; see also Cousins, 2007; Kepe, 2008; Metcalfe & Kepe, 2008; Okoth-Ogendo, 2003; Shipton, 2007, 2009, 2010; Sjaastad & Cousins, 2008). It is simplistic, however, to think that this conception of land is merely Polanyian. There are many overlaps, of course, but that could simply mean that Karl Polanyi's writings were influenced by his study of African societies (see, for example, Polanyi, 1957) where concepts of totemism and animism, for example, stress the interconnections between humans and nature, economy and environment, and religion and reality whereas current so-called environmentalism tends to adopt a more separatist, "we-humans-against-the-environment-and-animals-approach" (Adams & Mulligan, 2003; Hiedanpää & Bromley, 2016; Langton, 2003; Maathai, 2004, 2011). In turn, this Western Left Consensus neglects the interrelationships between environment and economy (Bromley, 1991; Hiedanpää & Bromley, 2016), space, society, and time (Obeng-Odoom & Bromley, 2020). This holistic Africanist conception of land also helps to decolonize nature from being hijacked by conservationists or "green grabbers" who offer a separatist view of economy and nature, leading to patronizing and ahistorical "concerns" about "extractivism," while simultaneously deprovincializing the idea of the commons (see, for example, Adams & Mulligan, 2003; Fairhead et al., 2012; Showers, 2014).

This strong emphasis on *the centrality of land* differentiates this book's conception of the commons from other methodologies. For example, this conception differs from the notion of "shared societies" developed by the Club de Madrid or from P2P production developed by Michel Bauwens and others, which are echoed in Figure 1.1. These other approaches leave out land entirely (for a discussion, see Prato, 2014; also see Bauwens et al., 2019) or consider land to be ordinary (e.g., P2P production advocates, together with other traditions captured in Figure 1.1, give no *special* place to land; for a discussion, see Bauwens et al., 2019, and Papadimitropoulos, 2018). For this study, however, land is fundamental and, hence, is emphasized throughout this book.

With this conception of land, the book also develops alternatives to analytical positivism, an approach that commonly interprets the commons as unchanging, an unfailing nod to socialism. For example, analytical positivism falters badly when looked at in the literalist approach to judicial interpretation used by some courts in Africa. Such courts follow the European technique of hinging modern decisions purely on dated precedents when current customs are rapidly changing (Asante, 1975, pp. xiii–xxvi, 1–26; Date-Bah, 2015, pp. 18–21), leading to bizarre outcomes. Outside the courts, analytical positivism has blinded many anthropologists, causing them to stick to notions of "culture" that were socially constructed by Europeans or have long been abandoned. While the courts take hasty flights based on precedents into wild generalizations, the anthropologists are often trapped into narrow interpretations of culture (for a fuller discussion, see Akiwumi, 2017). This book adopts a more dynamic approach. I carefully analyse modern conditions and, based on empirical evidence, draw inferences. The analyses in the book are a synthesis of the thesis of universalism and the anthropological antithesis of relativist interpretations of the concept of land.

My notion of the commons as land is also global, not just national, which tends to be the primary scale of analysis of both the Conventional Wisdom and the Western Left Consensus. At this global scale, the emphasis on the commons as land calls attention to imperialism, including the creation and maintenance of colonies and monopolies (Cobb, 2016, pp. 268–269), and highlights the tensions and contradictions about the Scramble for Africa. Africa is a commons, in this sense, for the taking of the rich, by force or fraud. This conception also enables us to see through the intricate ways that institutional processes become "normalized" or naturalized. This global aspect of the commons is linked to the local commons by issues of power and citizenship, as these have, over the centuries, been tied to processes of land ownership and the control of territory.

By definition, all commons cease to be so if they are commodified. When land, in particular, is commodified, not only does it cease to be a common, but this "commodified land" and the process of commodifying land also create social structures that shape society, economy, and nature adversely. Labour – but also citizens more widely (especially those in Africa and the Global South more directly dependent on land) – are harmed as a result.

Similarly, commoning land has a rather different potential: It can trigger new structures and processes that, in turn, can transform society, the economy, and ecology. Suggestions to make the planet sustainable through degrowth abound. They tend to emphasize individual responsibility for reducing consumption, enhancing market socialism, or promoting state socialism.

Paradoxically, such Western Left Consensus seeks to transform neither wage labour, the nature of money, the contents of markets, and TNC-based profits, nor the nature of the state, although they are institutions that support the relentless drive for growth (Exner, 2014, 2021; Toivanen & Kröger, 2019). More fundamentally, this Western Left Consensus calls for degrowth and the end of capitalism generally (Research & Degrowth, 2010). Yet many in this movement ignore the element of rent, private property in land as its core vehicle, and bonded labour as one core consequence (Obeng-Odoom, 2018, 2020). The commoning of land, on the other hand, provides concrete grounds and a path for firm steps to be taken towards correcting present and historical inequities, degrowth, and a steady state economy (for a detailed discussion, see chapters 3, 6, and 7; also see Daly et al., 1994, and van Griethuysen, 2012).

This commons-based mode of production is particularly promising because the emphasis on land is not limited to certain fixed communities (e.g., the commons that Ostrom studied), areas, or projects. Instead, it constitutes the grounds for a new society in which labour is liberated, the impulse for land-based money creation is transformed (in ways that could be consistent with the demands of the demonetization movement – see Exner, 2014), monopolies that constantly chase after profits are abolished, and rentier landlordism is curtailed (Cobb, 2019).

Markets continue to exist, but they are transformed into socioecological markets, which are supportive of gifts, solidarity, and exchange rather than seeking one vision of exchange for all (see Exner, 2014, pp. 13–15, 2021; Obeng-Odoom, 2021).

Thus, this book (1) engages both the Conventional Wisdom and the Western Left Consensus approach, (2) problematizes them, and (3) seeks to transcend them. This approach informs the choice of case studies in this book. As suggested in Figure 1.1, cities *must be considered* commons because they are commonly created and shared around urban land (Foster, 2016; Haila, 2011, 2016; Iaione, 2016; Obeng-Odoom, 2010a, 2020) or, because city residents have what Henry George (1883/1966) calls "the common rights to the soil" (p. 238). The Conventional Wisdom posits that the city is an open range system and, hence, a "tragedy" awaits the city that does not privatize its services and spaces (contrast with Rose, 1986). This view, in essence, drives the race to build "shiny new cities" to impose order in a development frontier (Côté-Roy & Moser, 2019). In turn, systematic attempts to privatize the city have become part of urban governance, drawing the ire of a movement of scholars and activists who organize around Henri Lefebvre's idea of "the right to the city" (Harvey, 2012; Webb, 2017). Curiously, aspects of the Conventional Wisdom appear – superficially, at least – to support the parallel existence of informal settlements in ways that could be germane to the Western Left Consensus characterized by its talismanic embrace of omni-commons (Bollier et al., 2015, pp. 258–270).

Technology, as well as the knowledge it helps to produce, could be regarded as commons too. Examples of such technological commons is Wikipedia and other open access publications, widely advocated by the open source movement, P2P group, and copyleft activists (Bauwens et al., 2019; Dugger, 2016; Kelly, 1981; Niman, 2011; Papadimitropoulos, 2018). Technology and knowledge are commons not just because they *ought to be* commons (although that emphasis is also important) but also because they *are* commons by reason of how they are created, the treatment to which they are subjected by mainstream economists, and the effects their adoption have on society, economy, and environment. The creation of technology and knowledge tends to be collective, and collective effort tends to enhance them. The idea that technology and knowledge are commons can also be traced to the argument that only their privatization, that is, patenting, can ensure they will avoid the "tragedy of the commons." For these reasons, Wikipedia is often regarded as a commons (see, for example, Euler, 2016).

Oil and water need little or no justification as commons and have typically been central to the commons debate. Questions about overproduction of, and the most efficient ways to extract, oil from its field have long been related to questions of the commons (Balthrop, 2012; Kim & Mahoney, 2002; Libecap & Smith, 2001; Libecap & Wiggins, 1984). Similar issues have been raised about water (e.g., Ingold, 2018; Ostrom, 1990; Sharma, 2012; Theesfeld, 2019). As "free gifts of nature," these resources have historically been at the very heart of the contentions about the commons.

The point about choosing these for case studies is not so much that I consider all of them commons, but rather that they are considered as commons by writers in both camps: Conventional Wisdom and Western Left Consensus. Consequently, they are systematically analysed as such: cities in chapter 4, technology in chapter 5, oil in chapter 6, and water in chapter 7. On these case studies and their interconnections, I throw the theoretical floodlights of my third way, the Radical Alternative, to demonstrate the myth of privatizing nature and to systematically develop a defence of prosperity and posterity through common landed property. In this way, the book avoids the triple problem of "fuzzy concepts, scanty evidence, [and] policy distance" that Ann Markusen (2003a, p. 701) has famously argued are the key problems of using case studies. Indeed, Markusen (2003a, 2003b) argues that much of (critical) social sciences today is reliant on singular, atheoritical, and poorly contextualized case studies. This book, on the other hand, tries to provide more comprehensive and more rigorous analyses than the standards associated with both the Conventional Wisdom and the Western Left Consensus research. By systematically engaging existing major debates, this book also tries to be more dialectical in its scope and, following Markusen's (2003a, 2003b) passionate plea for social scientists to be even more open minded about sources of data, be more encompassing in its range of data sources.

Sources of Data

This book draws on my study of land economy in Africa since 2001. While I draw on these years of study and research experiences to historicize and to contextualize, my data are more recent. They are from four years of field research I have conducted in West Africa, a synthesis of fieldwork data collected by others, numeric and non-numeric evidence from official surveys and civil society publications, analysis of opinions delivered by the courts, and detailed engagement with published data from, among others, original historical publications and statistical compendiums.

The existing practice of using only so-called scientific data has done much injustice to the accounts of the colonized. Data presented in travel books, in newspapers, in stories, songs, and poems have been dismissed. Even anthropologists, who, by the nature of their approaches are supposed to be more sensitive to the multiplicity of sources of data, have curiously denigrated anything that is not scientific ethnography. This practice continues to this day, although this failing has long been noted by leading anthropologists who pioneered the use of ethnography. As Mary Louise Pratt (1990, p. 27) noted in her contribution to *Writing Culture: The Poetics and Politics of Ethnography*, the discipline's most respected grammar of ethnography:

> The statement is symptomatic of a well-established habit among ethnographers of defining ethnographic writing over and against older, less specialized genres, such as travel books, personal memoirs, journalism, and accounts by missionaries, settlers, colonial officials, and the like. Although it will not supplant these genres altogether, professional ethnography, it is understood, will usurp their authority and correct their abuses. In almost any ethnography dull-looking figures called "mere travelers" or "casual observers" show up from time to time, only to have their superficial perceptions either corrected or corroborated by the serious scientist.

The complicity of this practice in colonial claims of "discovery" is obvious enough: dismissing the daily observations of Africans. Recognizing and engaging a wide range of data sources they and others produce is essential for building a more scientific, more holistic, and more decolonial analysis. Using a diversity of data sources is a strength. No one source of data can provide the holistic coverage needed for analysing the commons. Interview data are important, as textual data, but not in isolation. Economists who rely solely on analysing large datasets risk losing the meanings that intimate knowledge of particular cases and context can provide. Those analysing only texts or macro world systems may lose the micro and micro-macro insights of sliding between relativist and universalist sources of data. As George Bob-Milliar (2020) has noted, overcoming colonial methodologies must entail not just how we study what but also who does the research. Reflecting on solutions elsewhere is important but, ultimately, emphasizing contextual alternatives must be given far more credit than existing research practice offers. As a Southern scholar based in the Global North, I am uniquely placed to attempt to dissolve the "outsider/insider" tension. Yet I subject my work to the scrutiny of peers whether in the South or the North, or multi-scalar, as I am.

Some of the data have been peer reviewed in earlier publications. Others respond to questions raised at the end of earlier data collected. The rest fill in gaps in existing studies. Judicial opinions are public, but only their media versions tend to be read; I base my analysis on a combination of legal analysis and the original opinions collected directly from judiciary sources. Court decisions constitute a rich data pool for political-economic analysis, especially because law and its reforms have been central to the commodification of the commons (see, for example, chapter 8). Simultaneously, law also provides another important avenue to protect the commons (see, for example, chapters 6 and 8).

What role is played by law, then, depends on wider political-economic factors. Some of these are the interests of those who finance legal reform (Manji, 2013), lawyers and their associations (Gould, 2006), and the socialization of judges within the wider judicial system (Date-Bah, 2015). In using legal sources, therefore, I have taken seriously the context (colonial, neocolonial, neoliberal, and modern imperialism, for example) within which decisions are given (Moore, 1986), the prevailing legal techniques for interpretation, and the changing nature of the context (Asante, 1975; Date-Bah, 2015). Judicial opinions, in this sense, are not merely statements of law; they offer a glimpse into what leading institutional economist J. R. Commons (1924) called *The Legal Foundations of Capitalism* in his book by the same name.

For this book, however, I also collected judicial opinions that enable capitalism but also constrain it. Likewise, the nature of law gives some insights into colonial mindsets and helps to see the braided links between colonialism and neocolonialism on the one hand, and capitalism and imperialism on the other (see Ince, 2014; Manji, 2013; and chapter 7). This multilayered evidence base provides the book with a transparent and strong backbone. As with other social science books, this book demonstrates the interrelations in data presented as stand-alone sets and enables me to extend analyses, which I cannot do in short articles. The use of this eclectic mix of data is indicative of the pluralist and transdisciplinary nature of the book, not in terms of merely aggregating small steps along the way but, rather, in developing new arguments.

The Arguments

Drawing on these data and the radical approach, this book makes three arguments: First, the Conventional Wisdom about the commons is highly unsatisfactory. Treating the commons as private property (Garrett Hardin) is problematic but so is treating it as a common-pool resource (Elinor Ostrom). The implementation of policies patterned after both conceptions have created major "social costs," that is, "all direct and indirect losses suffered by third persons or the general public as a result of private economic activities" related to the privatization of nature (Kapp, 1971, p. 13) in ways that contradict the promises of the Conventional Wisdom. So, if there are "transaction costs," supposedly generated by the existence of widespread commoning, privatizing nature generates widespread social costs, which are not merely "short-run price paid for a high level of long-run efficiency and social performance of the economic system" (Kapp, 1971, p. 15), as has often been claimed by Conventional Wisdom. Rather, these are complex costs.

Dialectically related to progress, these costs are cumulative and generate *continuing, often worsening* dynamic inequalities. These forms of stratification for different and differential races, classes, and gender are not, and cannot, be fully accounted for by a mere mainstream microeconomics cost-benefit analysis based on static, individualized costs (see Argyrous, 2017; Kapp, 1971, pp. xxiii–xxiv; Obeng-Odoom, 2020; Stilwell, 1999). These social costs are not just generated and maintained within nations and between nations; they are also produced and reproduced at different scales. So, they entail pre-colonial, colonial, and neo-colonial costs, as well as the costs of imperialism.

Second, the Western Left Consensus tries to provide a response to these problems by creating a more transformative view of the commons and commonizing the product of labour or capital (e.g., making technology a commons). This "remedy,"however, could create additional problems because it would generate social rents that are privately appropriated in a way that would make uneven development structural. Some advocates of the Western Left Consensus (e.g., Cato & North, 2016) attempt to common the land, but their approach of physical redistribution of plots of land – underpinned by the notion of "equal factors of production" – cannot redress the problem, which is then compounded by their neglect of the difficulties of development and underdevelopment at a global scale.

Third, commoning land in the ways developed by my third way, the Radical Alternative, would address these problems locally and globally, and it would also prevent them from happening. More fundamentally, the Radical Alternative could lay the foundations for prosperity without destructive growth (Gordon Nembhard, 2014a, 2014b). In this sense, making land the most fundamental of all commons "redirects the economy toward community, the environment, and a sustainable future" (Daly et al., 1994). Thus, analytically, this book solves the holy grail in commons debate: how to link ecology with technology and social relations in general, and how to combine either land, technology, and social relations beyond a consensus approach. Bringing these dimensions and intersections together, while demonstrating the centrality of social-nature relations (land) is an important breakthrough.

The Structure of the Book

These arguments are developed in the four parts of this book, which begins with Part A: The Problem, dealing with planetary socioecological crises. Part B considers how these socioecological crises have evolved over time, including showing major analytical difficulties and unsettled questions, while providing a new approach to rethinking the terms of the debate. Then, on the basis of this rethinking, Part C empirically shows that although privatizing the commons creates major problems (thus contradicting Hardin), the reformist solution – based on common-pool resource thinking (Elinor Ostrom) – is unsatisfactory and creates even more difficulties. More fundamentally, the book shows that the Western Left Consensus, the so-called alternatives of commoning (centred on commoning the fruits of labour and capital), is similarly problematic. As brought together in Part D, commoning land and adopting the conceptual approach of this book would solve these problems and could create the conditions for prosperity without poverty and develop the foundations of a socioecologically just society.

Chapter Overviews

Chapter 2: Historical Debates on the Commons

What is the historical context in which the controversies about the commons should be placed? In contrast to the compression and natural history of the Conventional Wisdom, which leads to the teleologically problematic claim that private property in the commons *naturally* arises *over time* from backward commons property, chapter 2 adopts a long-range and materialist conception of history. It does so by situating the current debates about the transformation of the commons within historical parallels. These focus on the processes of creating property in the commons. They go beyond the unidirectional, Westernized story of the "enclosure." Instead, these accounts describe how commons debates are linked to the emergence of money as a social relation and how the debaters have approached the topic. This historical analysis is particularly important because, although not systematically embraced, it can provide stronger foundations for the current analyses of the commons. Three insights from the chapter illustrate this modest contribution. First, between the third and the nineteenth century, debates about whether the effectiveness of the commons as a mode of organization were heated, but since the twentieth century, it has become increasingly accepted that the commons are problematic. That transition was influenced by political-economic interests rather than by scientific ones and the proof or mere effluxion of time. Second, the commons debate was also central to the debate about the emergence of money, meaning land and its commodification have historically been tied to the development of money. Third, with the growing acceptance of the idea of a "tragedy of the commons" has come a growing inferiorization of land tenure systems in the Global South. As this dominant paradigm for evaluating the commons is politically biased, it is necessary to rethink the very framework of commons research.

Chapter 3: Rethinking the Commons

Chapter 3 accepts that rethinking challenge. It does so by invoking the work of Ostrom, which has convinced mainstream economists that collective governance of the commons is possible without the attendant challenges of the tragedy of the commons and free-rider problems. However, the chapter argues that a more systematic appraisal of Ostrom's work shows that it is hardly an avatar of the commons and society. Ostrom's work contains no concept of justice – a central pillar for appreciating and addressing socioecological crises. Rights are commonly mentioned in her work, of course, but her idea of rights is extremely limited, often tied to the notion of joint, rather than equal, rights. For Ostrom, the notion of the commons is socially separatist, partially economic, and environmental but not ecological (which is also environmental but embodies social costs, the questions of justice, and the processes of time and space). Ostrom's analysis of the commons is commendably historical, in parts, but it is not systematically so and, hence, her proposed "collective action" to save the commons provides fertile grounds to grow the real threats to the commons.

A strikingly different and more holistic approach to the commons is offered by African institutions and practices. These are closest to the ideas offered by Henry George, who posits the commons as *the* most important path to social, economic, and ecological sustainability. Unlike Ostrom who studied the commons as a "scientist" desirous of showing "scientifically" that there is another good that is neither private, public, nor club-based, George studied the commons to understand and remove injustice at the roots. In turn, his approach is more critical and certainly more relevant today in showing that another world is possible, even if both historical and contemporary examples of the commons and how they function(ed) suggest that George's work, too, requires significant changes to update its framing of the meanings, prospects, and future of the commons, particularly in Africa. For example, it should include ideas from "the embedded approach" of Karl Polanyi. The postcolonial analysis of Frantz Fanon and others should be embraced, too. This rethinking raises many questions, including the specific ramifications of the commons for society, economy, and environment. Answers could be provided by case studies about cities (chapter 4), technology (chapter 5), oil (chapter 6), and water (chapter 7).

Chapter 4: Cities

The ecological crises that threaten the nature and future of cities are often blamed on the "fact" that cities are open range spaces and are ungoverned, or ungovernable, especially as they become larger and larger with population growth. This urban "tragedy of the commons" was presumably challenged by Elinor Ostrom. In principle, however, what Ostrom questioned was whether market fixes, technology, or privatization is the only way. She did not contend that they are ill-advised as institutions. Indeed, in her approach to the urban commons, Ostrom proposes that all these instruments of conventional thinking could be part of the solution. Urban common-pool resources, namely, informal communities, slum settlements, and gated estates should complement the picture. These urban commons, she contended, arise from individual rational decisions to escape top-down urban planning. Polycentricity, then, was Ostrom's approach to the urban commons and its crises. The seeming consensus in this Conventional Wisdom overlaps with Western Left Consensus on the commons and its near-total focus on neoliberalism as the key antagonist. So, while the Western Left Consensus excoriates the state for being an agent of neoliberalism, the Conventional Wisdom critiques the state for arbitrary decisions that lead to the formation of slums.

These claims are well known. The question chapter 4 investigates is not what the claims are but whether they are borne out by empirical evidence. In doing so, the chapter investigates the drivers of the so-called urban common-pool resources, the conditions of people in these communities, and the extent to which they contribute a lasting solution to the urban socioecological crisis alongside the claims about marketization, privatization, and technological diffusion. The data for the analysis include material collected first-hand in cities in Africa.

The evidence calls into question both the Conventional Wisdom and the Western Left Consensus on the urban commons. Informal urban common pools or communities remain widespread, but they have been produced by more structural processes that include but transcend urban neoliberalism. Indeed, neoliberalism could also be regarded as an effect of longer processes of colonialism, neocolonialism, and bigger issues of modern global imperialism. When dealing with dire conditions of life, especially difficult work conditions, the Ostromian thesis that informal economies are bright spots of liberation is questionable. The contribution of such communities to resolving the ecological crises in cities is notable, but it is structurally limited. Privatization might have its place in society, of course, but markets have not deterred polluting behaviour. If anything, they have augmented it. If there is a tragedy, it arises from the privatization of nature generates waste pollution. The privatization of nature could also lead to serious emissions in cities through monopolistic extractive industry practices. As polycentricity ignores these structural processes, Ostrom's contribution to addressing urban problems is severely limited, while the contribution of Western Left Consensus is partial at best.

By linking ecological questions to labour and waste and, hence, to the different types of value in land, labour, capital, and waste, this chapter contributes to moving forward the analytical literature currently centred on labour, capital, and waste. The chapter also tries to move the analytical literature beyond the mere "neoliberalization of nature," where it is currently stuck (Castree, 2008a, 2008b). Instead, I probe the problematic relationship between neocolonial and neoliberal forces of marketizing the urban commons and the deepening urban socioecological crises in the urban commons. Whether recent technological advances constitute a panacea requires further analysis.

Chapter 5: Technology

The widespread adoption of technology in Africa and the Global South more generally provides another opportunity to reassess Conventional Wisdom. According to this paradigm, technologies remove the limits to growth, address labour problems, and, more fundamentally, bring about global income and wealth convergence while addressing socioecological crises. If access to such technologies is restricted, the Conventional Wisdom holds, such technological gains could improve substantially because innovators would have both necessary and sufficient incentives to make scientific breakthroughs.

Existing critiques of these growth models from the perspective of Western Left Consensus question their triumphalism, problematizing the existing digital divide and the devastation that more enclosure might cause. Indeed, Western Left Consensus points out that technological change has led only to limited gains in the ability of cities in the Global South to access the fruits of global production. These gains

are, however, coupled with problematic downgrading of local industries and the rise of new forms of economic dependencies. Yet these
problems could be addressed if technology were made appropriate to
the South, for example, by bringing it under the control of workers and
citizens more generally.

Although compelling, much like the Conventional Wisdom and its
growth models, these critiques frame the technology question only in
terms of labour and capital. Evidence from cities in the Global South,
analysed with the aid of Georgist political-economic lenses, however,
shows that such framing is problematic. Inherent in technological
change has been the rapid increase in urban land rent driven largely by
technologically mediated speculation. This dynamic could have corrosive implications for real wages, which would tend to decline over time
as more rent or interest is paid. This problematique also drives uneven
development, which is produced, among others things, by actions
and inactions to enhance speculative rent extraction. Combined, these
effects could make growth even more fragile, inequality even more
structural, and socioecological crises even more complex regardless of
whether technology becomes a common.

Chapter 6: Oil

It is particularly useful to study oil because its nature, effects, and governance have always entailed land reform. Indeed, with oil regarded
as land, the colonial appropriation of African oil was socioecologically constructed as an attempt to address the tragedy of the commons. The colonizers sought to "save Africans from themselves" and
to save the world by seeking to govern oil the oil fields, which, much
like other fields, they considered *terra nullius*. Colonial and corporate interests were more blatantly yoked together in that era. Under
neocolonial imperialism, however, this braided link is framed as
non-existent. Instead, the Conventional Wisdom claims that the transnational corporation, as an independent and separate entity, pursues
oil "unitization."

According to advocates of unitization, overlapping landed interests in oil fields, that is, oil common pools, could lead to suboptimal
extraction practices, which could generate several problems typically
discussed as *resource curse*. It is better, then, to unitize or bring all the
competing interests under one oil lease held by one oil producer with
experience to develop oil "sustainably." Western Left Consensus, on the
other hand, takes the view that no oil should be drilled, insisting that
posterity is better off if oil were left in the ground.

The positions of the Conventional Wisdom and the Western Left Consensus raise the following questions: should the oil field be monopolized and run by an enlightened TNC with considerable experience or, for fear of the tragedy, should oil not be drilled at all? Chapter 6 demonstrates that commoning oil through a strategy of energy sovereignty is a more effective strategy and more consistent with the aspirations of historical Indigenous and black protests against private property relations in oil.

The case study of oil shows that, although challenging, a strategy that de-emphasizes economic growth and stresses autonomy, distribution, and energy sovereignty could be workable and superior to both the Conventional Wisdom and the Western Left Consensus, which, in practice, entails a missionary agenda to dictate how African countries can best govern their oil commons. Indeed, although apparently more consistent with Indigenous and black struggles against oil, the Western Left Consensus misunderstands the nature of Indigenous and black protests. Their perceived "solidarity" is not only elitist but also merely rhetorical.

This lesson, centred on the dangers of existing threats to the commons and the prospects of an alternative path to commoning, is not peculiar to oil. Demonstrating that it is widespread, by using the last case study, the water commons, is the task of chapter 7.

Chapter 7: Water

The recent surge in the marketization of the commons in Africa – especially of water bodies – warrants careful political-economic analysis. Three questions remain intractable: (1) Were there markets in the beginning? If so, how have they transformed and, if not, how did markets arise and evolve? (2) What are the outcomes of such markets for people, their livelihoods, and their environment? (3) How should we interpret the outcomes of water markets and should water should be commodified at all? For advocates of Conventional Wisdom, water markets have arisen because of the inferior nature of Indigenous or customary systems, which are incapable of offering precisely what water markets offer Africa: economic and ecological fortunes that traditional modes of governance do not. Western Left Consensus seeks the commonizing of water but mainly on the basis that it is a free gift of nature privatized since the era of neoliberalism. Chapter 7, on the other hand, investigates the social history of marketization of the commons and probes the effects of marketization in terms of absolute, relative, and differential/

congruent outcomes, as well as the opportunity cost of the current water property rights regime.

The empirical evidence shows that markets have been socially created through imposed and directed efforts. Some jobs have been created through investment, but such employment is not unique to marketization and private investment. Indeed, the private model of property rights has worsened the distribution of water resources not only within different property relations in Africa but also between diverse property relations. Water markets have been responsible for much displacement and trouble for communities and for nature. Overall, there is no necessary congruence between the promises made by advocates of the Conventional Wisdom and how communities experience water markets. In contrast to the Western Left Consensus and its causal theory of neoliberalism, the commodification of water has a much longer history. Tighter state regulations for the use of inland and transboundary water sources might temporarily halt the displacement of communities sparked by marketization of the commons, but only one fundamental change can guarantee community well-being: to regard access to and community control of water as constitutionally sanctioned human rights and as *res communis*. The details of this Radical Alternative of res communis, however, are developed more fully in chapter 8.

Chapter 8: Concluding Remarks: Towards a New Ecological Political Economy

After the completion of the proof from chapters 4 to 7, the book concludes with both reflections and contemplations for the future. Chapter 8 highlights the principal arguments made in chapters 1 to 7, shows both the impediments and the prospects for change in terms of policy and political action, and draws out key conceptual and analytical lessons for the future. Written as "emancipatory social science" (Wright, 2010), chapter 8 provides the outlines of a new ecological political economy that seeks to address the analytical gap in much of the writing on "reclaiming the commons" currently focused on political action. This new ecological political economy, on the other hand, develops concepts such as *rent theft* and *just land*. As a stepping stone, this approach opens the door to a more compelling analysis of the commons within and beyond the "age of uncertainty."

PART B

The Debates and a Path through Them

Chapter Two

Historical Debates on the Commons

Introduction

How did private property emerge? In what ways did private property spread? What analytical paradigms have helped to address such questions?

These questions are central to commons research, but they are often poorly treated, usually in isolation but also together. Typically, apart from repeating Karl Marx's account of how the enclosure of English commons paved the way from feudalism to capitalism, they are usually overlooked. Even Anne Haila, a prominent scholar in this field, deals with "debates on genealogies" (see, for example, Haila, 2016, p. 23) but not with the other questions. Neither Garrett Hardin, Elinor Ostrom, nor advocates of the Western Left Consensus systematically engage these questions. The authors of *The Open Fields* (Orwin & Orwin, 1967) attempt to address these questions, but they start their enquiry in the nineteenth century and restrict their investigations to rural England. Andro Linklater (2013) provides a *geographically* more diverse account in *Owning the Earth*, but his story starts only in the sixteenth century. He also excludes a systematic appraisal of analytical paradigms and neglects to consider the centrality of money (two key strands in the history of the commons) to his "transforming history of land ownership." The highly influential *Encyclopedia of Political Economy* claims that "the common property 'problem' was originally articulated best by Hardin" (Dragun, 2001, p. 119), but it provides no systematic historical basis for the claim. Others such as Derek Wall (2014) blend a so-called Hardin account and that of Karl Marx, but is this a more satisfactory approach?

Economists have offered three criticisms. First, they have sometimes claimed that Hardin misused the term *commons* when in fact he was referring to *open access* resources. Second, they have tended to claim that Hardin did not argue for privatization only. Instead, he argued for either private or state ownership (for a review, see Cobb, 2016). Third is the criticism that the existing approach to stating the debates on the commons compresses the history of the debates (see, for instance, Espin-Sanchez, 2015). To correct history, according to these critics, it is crucial to focus more on the contributions of new institutional economists such as Mancur Olson, Armen Alchian, Harold Demsetz, and Ronald Coase and take concepts such as transaction costs seriously. It can also be argued that the literature is too Eurocentric, often paying relatively little attention to research on the commons in the Global South (Sjaastad & Cousins, 2008).

Some of these criticisms are welcome because they invite further thinking on how the history of commons debates could be made more comprehensive, even if doing so is intellectually demanding. Others can be addressed easily. What Hardin meant by the "commons," for instance, can be understood by studying his 1968 article closely. There, Hardin recognized that the state can be a manager, but he concludes that it is an incompetent manager compared to the market.

Accordingly, he recommends greater enclosures and marketization. The suggestion that the problems in Hardin's claims were merely the result of linguistic or nomenclature slippage is similarly easily addressed. C. S. Orwin's (1938) historical research shows that the commons were also called "open fields." And Orwin's co-authored book *The Open Fields* (Orwin & Orwin, 1967) was reviewed by Joan Thirsk (1964) as a contribution to the commons literature. The issue with the commons, then, relates more to the political economy of property. Scholars such as Hastings Okoth-Ogendo, Parker Shipton, Benjamin Cousins, and Thembela Kepe (e.g., Cousins, 2007; Kepe, 2008; Metcalfe & Kepe, 2008; Okoth-Ogendo, 2003; Shipton, 2007, 2009, 2010; Sjaastad & Cousins, 2008) have enriched our knowledge of these debates by offering detailed analyses of how colonial, neocolonial, neoliberal, and imperial forces combine to transform the commons in Africa and the Global South more generally. They show how ideas of titling have led to the transfer of land from peasants to bankers in Africa. Focusing on contesting ideas propagated by economists such as Hernando de Soto and, more generally, by the international development agencies, this body of work throws light on the external forces behind the privatization of the commons. John Pullen (2013) has also offered micro histories of how various scholars contributed to the commons debates.

The demand for a comprehensive account, however, requires a more careful response that includes, but also transcends, the story of new institutional economics centred on the natural evolution of common to private property because of the superiority of the latter. If the account is to be decentred and diffused, then the narrow focus on the internal threats to the commons (e.g., Alchian & Demsetz, 1973) or on the proximate external drivers (e.g., colonialism) often emphasized by new institutional economists such as Daron Acemoglu (e.g., Acemoglu & Robinson, 2013; Acemoglu & Verdier, 1998) and postcolonial writers must be widened. Within this new net can be found pre-colonial and external imperial threats to the commons and how they intermingle with internal tensions and contradictions (see Showers, 2014).

In seeking to provide these missing building blocks, this chapter pays less attention to what has already been emphasized in the literature (the contributions of Marx and Hardin, for instance) and, instead, highlights the pre-eighteenth-century debates and extends the literature that clarifies the eighteenth and post-eighteenth-century debates (see, for example, Haila, 2016, pp. 26–45). The chapter also shows how those debates have evolved and, hence, shows the nature of the twenty-first century debates. This evolution, the chapter demonstrates, intermingles with the shift from classical economics through neoclassical economics and (new) institutional economics, to the Western Left Consensus.

The chapter shows that, neither Hardin's widely quoted paper nor Karl Marx's account over enclosures does justice to the complex and detailed history of the commons. That rich history, going back as far as the third century, the heated debates, and the prevailing analytical approaches reveal three key themes.

First, the issue of the best form for property regimes, which was widely contested in the pre-twentieth-century era, became surprisingly widely accepted from the twentieth century on, much like the Kuhnian shift from classical to neoclassical economics. As with Kuhnian as opposed to Popperian shifts, however (for a general discussion of Kuhn and Popper, see de Vroey, 1975), these scientific revolutions were not mere advances but highly political victories for the Conventional Wisdom, legitimized by the many Nobel Prizes awards to its key advocates. Second, the debates in the literature emphasize the centrality of property to the emergence and continuing importance of money. Third, the debates on the commons have tended to inferiorize the land tenure systems and property regimes in the Global South. So, the Global South has increasingly come under pressure to reform its institutions into the image of the West.

The rest of the chapter is divided into four sections. The first, "Money, Debt, and the Origins of Private Property," shows the centrality of property to the emergence of money. This important focus sheds light on the emergence of private property from the commons; helps to highlight the need to engage questions of money in the analysis of the commons;

and supplies the context for the discussion on the nature of Indigenous economies, the place of barter, and the contemporary interest in pursuing degrowth in chapters 6, 7, and 8.

"The Spread of Private Empires and Private Property: Religion" focuses on how economic interests and power groups developed ideological theories about privatizing property to justify their own interests. This section of the chapter helps to clarify that the debates about property, while sometimes dry, are not merely idealist or academic. Rather, they tend to be seriously materialist. So, any attempt to revise the grammar of property must also be prepared to confront the so-called "property lobby" (Haila, 2016). The third section, "The Spread of Private Property: Markets," analyses the marketization of land through the insidious spread of land title registration. The chapter ends with the section "Privatizing the Commons: Competing Analytical Paradigms," which highlights the golden age of the new institutional economics approach, which culminates in the birth of Elinor Ostrom's Bloomington School of New Institutionalism (see Aligica & Tarko, 2012, p. 237). This new institutional economics approach emphasizes the common-pool approach, and the Nobel recognition of it brought a new visibility to the Conventional Wisdom under which the Western Left Consensus seeks to light its own path.

Money, Debt, and the Origins of Private Property

The emergence of private landed property is closely tied to the use of the commons as security for credit, the rise of money to defray such debt, and the role of money in the economy. The story of debt has a long history, the first five thousand years of which has been detailed by David Graeber (2011). The debates here can be quite complex. They are often centred on the emergence of money *ab initio* (over which quite separate controversies exist, including making barter more efficient) and new uses of money (where money already existed). Debt also existed in distinctive forms in pre-colonial Africa, as discussed in chapter 7. For the current analysis, how money and debt are linked to the emergence of private property requires clarification to help contextualize the debate on the emergence of the commons.

The Lockean and neoclassical economics idea that the acquisition of private property is merely the product of hard work and the result of free exchange is a useful starting point because it is widely taught (Bell et al., 2004; Lea, 1994; Obeng-Odoom, 2017b, 2019b; Ryan-Collins et al., 2017; Theobald, 1997). By this view, money emerged to assist the free exchange of land, which was cumbersome under the barter system. Characterized by qualities such as portability, durability, and acceptability, money became the medium of exchange, a store of value, and a unit of measure of value to facilitate the free exchange of land (Bell et al., 2004).

Existing systematic research (e.g., Lea, 1994; Ryan-Collins et al., 2017) suggests that this historical account is unsatisfactory. According to this body of scholarship, in Rome and Greece where much of the history has been well documented, power, imperialism, and unequal exchange were important features of the transformation. This line of analysis is interesting, but it is sketchy.

Could experiences in Africa help to clarify these processes? Historical research by Forstater (2005) shows that the colonizer introduced a land tax in some African colonies to force the Africans to hire themselves out for wages. As most Africans were not accepting of the market logic to work for wages and preferred instead to barter or till the land for subsistence – practices inconsistent with European views of "progress" – land tax and other direct taxes to be paid in the colonizers' local currency were levied. In turn, the Africans were coerced to offer their labour in exchange for money, much of which was used to pay off burdensome taxes. The colonizers had the money; what they needed was labour. To get it, they had to make the Africans need money.

A non-land tax such as an income tax could not serve this purpose, because the Africans could simply stay out of the cash economy, whereas the land tax could not be avoided. So desperate were the colonizers for African labour that they would burn down the houses of defaulters or force defaulters to watch the sun from sun rise to sun set (Forstater, 2005, pp. 59–60). To pay off such debt and other forms of it, Africans had to use credit, often secured by their land, which they lost when they defaulted (e.g., Shipton, 2007, 2009, 2010). Thus, money was introduced to help pay off debts, to measure the value of these debts, to facilitate the process of annexing land from the commons through land sales, and to enable debtors to take out more credit secured by their land.

Apart from the European and African experiences considered in this chapter, broadly similar dynamics have been recorded for the Americas and in the Middle East (Duchrow & Hinkelammert, 2004, pp. 5–7; George, 1898/1992, pp. 482–528; for a discussion of the contributions of Keynes and Commons, see Tymoigne, 2003; for the rest, see Decker, 2015). This system was held together by haute finance, or the international money system based on credit, trade, and perpetual accumulation, which enabled world powers to accumulate large parcels of land elsewhere through so-called fair trade (Polanyi, 1944/2001; Seccareccia & Correa, 2017).

The global economy was characterized by unequal exchange. Powerful landlords extracted the labour of the powerless who were caught in bondage. To extricate themselves, they had to solicit the help of their families with whom they worked for the continuing accumulation of the landlord (Duchrow & Hinkelammert, 2004, pp. 5–7). In this process, as stressed by Frank Decker (2015, p. 944), "the critical domain of the state is the maintenance of property law, the enforcement of debt contracts, and the provision of institutional arrangement establishing an effective lender of last resort." A fundamental process, this dynamic was complemented by many other practices that popularized the idea of private property.

The Spread of Private Property: Religion

Religion is one of such mechanism for popularizing private property. Marx took a dialectical view of religion. He once observed that "religious distress is at the same time the *expression* of real distress and also the *protest* against real distress. Religion is the sigh of the oppressed creature, the heart of a heartless world, just as it is the spirit of spiritless conditions. It is the opium of the people" (as cited in Chakrabarti et al., 2016, p. 339; emphasis in Chakrabarti et al.). Yet only "the opium of the people" part of the quotation is usually used. Accordingly, political economy tends to be materialist. Land economists, however, have usually considered religion carefully. It is an important source of insights into nature and the transformation of land. In principle, many religious teachings emphasize the imperative for land to be considered a commons. In practice, religion has also contributed to the spread of private property in land.

Take Islam. Much of its teachings are pro-commons. Consider three examples. First, speculation on, and concentration of, land are abhorred by the Holy Quran, which explicitly forbids *Riba* (rents or "over and beyond" one's fair share; see Behdad, 1989, p. 194). Second, workers are entitled to what they produce. Third, ultimate ownership of land is God's. As mere trustees of land, humans should avoid the incessant accumulation of landed property (Behdad, 1989; Razif et al., 2017; Zaman, 2019).

Attempts to realize these ends have been consistently frustrated by propertied interests. They created "controversy" about the true teachings of Islam. For example, some propertied interests have successfully created the alternative Lockean view that landlords capable of real estate investment could accumulate land. Indeed, speculation, typically frowned upon in Islamic economics and teachings, is now rationalized. The detailed history of the emergence of speculation in Islamic economic thought has recently been presented by Nor Fahimah Mohd Razif and colleagues (2017), so repeating it here is not necessary. What ought to be emphasized is that, today, speculative capitalism has come to characterize many Islamic property practices (see Razif et al., 2017). Of course, *waqf* land still exists as an alternative (Zaman, 2019), creating the historic parallel tension between materialism and secularism even within religion.

Similar comments apply to Christianity. Guy Shrubsole's (2019) book *Who Owns England?* and Brett Christophers's (2018) book *The New Enclosure: The Appropriation of Public Land in Neoliberal Britain* have systematically demonstrated the interlinkages across church, state, TNCs, and Crown in spreading the idea and social practice of private property. The state, along with the Crown and the church, uses discourses of (in)efficiency to systematically transfer significant parcels of common and public land to private interests. The process has been legitimized so silently that the great land transformation has taken place almost without notice. Excellent in their expositions and revelations and, hence, widely read and reviewed (see, for example, Dobeson, 2019), these accounts provide rich insights that are *particular* to England. As broad-ranging accounts (e.g., Cahill & McManon, 2010), they have successfully demonstrated the strong connections between land monopoly and capitalism centred on minerals such as coal (for example, Fine, 1990). However, they can be extended to throw light on the systematic intersectional account of the role of the church in spreading the idea of private property.

Indeed, as the role of religion in this process is both particular and general, it is important to dig deeper. The experience of the early Catholic Church is worth considering in this respect because of its sphere of influence, recently documented by Elizabeth Foster (2019) in *African Catholic: Decolonization and the Transformation of the Church*. Initially, the church provided sustained criticism of private and absolute property. In the third century and, hence, before the transformation of the church, the early Christian writers such as St. Augustine, St. Cyprian, and St. Chrysostom argued that private landed property was sinful. Land, the argument was made, *is* common property. Owning property in common, then, was said to be consistent with Scriptures and the good society.

According to Jamieson (2014, pp. 17–18), the early church and the Christian leaders, notably Jesus of Nazareth, took a firm stance against private property in land. The Old Testament, which Christ came to "fulfil," provided categorical statements against private property in land. Based on the lived experiences of oppression unleashed by private property in land, Moses was keen to institutionalize land as a commons (George, 1884). What Jesus did, then, was to admonish, to teach, to proclaim freedom from property bondage, and to endorse the celebration of the Jubilee Law of Jehovah God, which prohibited the sale of land because land belongs to God and all other people (of all faiths) are God's tenants. According to the Book of Leviticus in the Bible, God said:

> 23 The land must not be sold permanently, because the land is mine and you reside in my land as foreigners and strangers. 24 Throughout the land that you hold as a possession, you must provide for the redemption of the land. 25 If one of your fellow Israelites becomes poor and sells some of their property, their nearest relative is to come and redeem what they have sold. 26 If, however, there is no one to redeem it for them but later on they prosper and acquire sufficient means to redeem it themselves, 27 they are to determine the value for the years since they sold it and refund the balance to the one to whom they sold it; they can then go back to their own property. 28 But if they do not acquire the means to repay, what was sold will remain in the possession of the buyer until the Year of Jubilee. It will be returned in the Jubilee, and they can then go back to their property. (Bible Study Tools, n.d.)

These provisions forbade both the sale and purchase of land. If under conditions of hardship, the poor sold their land, their relatives had to help to redeem the property. Or if the poor became sufficiently prosperous to redeem the land, they had to do so themselves. If neither the relatives nor the poor were able to redeem the landed property, after 50 years, the property had to revert to the poor without compensation to the purchaser. These provisions were not "making property sales more like a leasehold transfer, with reversion to the permanent owners at the following Jubilee" (Small, 2004, p. 163). Instead, they constituted a bar to private property in land (Duchrow & Hinkelammert, 2004, pp. 21–22).

As Christianity gradually became the official religion of the Romans, the interest in spreading the idea of private property gained official acceptance. The Roman law of Dominium which, according to Jamieson (2104), is "the legalisation of property in land which had been taken by plunder and conquest" (p. 17) exerted a grip on the church and offered the theological direction for church position on landed property. As the church became a secular church, a church that went to war, and one that sought to expand its wealth, the church became, in the words of Terry Sullivan (2008, p. 19), "the church of the landlords." These landlords coerced their labourers to become members of the emperor church, called the Catholic Church, which, in turn, became a global landlord in the sense that it acquired land across the world, including large African estates (Sullivan, 2008, p. 46); much of the class of bishops was also the class of absentee landlords (Sullivan, 2008, p. 141).

Many in the emperor church wrote enthusiastically in favour of property in land. Alexander of Hales and Albert the Great, for example, argued a more natural school perspective. In this view, private property was accepted as the key motor for economic progress and, hence, it was encouraged. What was required of the lucky few, the private landlords, was to help the unlucky poor who had been dispossessed of their land. This view, according to Garrick Small (2004), was considered as reconciling the notion of private property of Milton Friedman with Adam Smith's (1776/2007) moral economics, which is resolutely against private property in land:

> As soon as the land of any country has all become private property, the landlords, like all other men, love to reap where they never sowed, and demand a rent even for its natural produce. The wood of the forest, the grass of the field, and all the natural fruits of the earth, which, when land was in common, cost the labourer only the trouble of gathering them, come, even to him, to have an additional price fixed upon them. He must then pay for the licence to gather them; and must give up to the landlord a portion of what his labour either collects or produces. This portion, or, what comes to the same thing, the price of this portion, constitutes the rent of land, and in the price of the greater part of commodities makes a third component part. (p. 43)

As Sullivan (2008) showed, this transformation, the shift from debarring both absolute and conditional property to celebrating the acceptance of conditional property, the condition being the expectation on the private landlord to help the disposed and the poor (Pullen, 2019; Small, 2004, 2013), started when Judas Iscariot was bribed to betray Jesus Christ. However, its effects became evident when Emperor Constantine took over the reign of the Roman Empire early in the fourth century. This change in stance, then, seemed to have been propelled by an interest to support the church and the landed property it had itself accumulated. The climax of these dramatic transformations was Pope Leo's famous encyclical *Rerum Novarium* (see a more detailed history in Pullen, 2019). Published in 1891, this letter by Leo XIII was widely interpreted as a critique of Henry George.

In contrast to the Georgist approach to land, which prioritized making and keeping land as commons, Pope Leo represented the official position of the church as neo-Lockian, putting forward arguments in favour of privatizing the commons, with the injunction on landlords to use their landed property for the common good of humankind. While Henry George systematically showed the limitations of this "theological individualism" (Cobb, 2016, p. 269) in *The Condition of Labor: An Open Letter to Pope Leo XIII* (George, 1891), this version of Roman Catholicism has nurtured "the founding principles of liberalism." Otherwise called "the stewardship role of individual property," this view stipulates "individual control and social benefit" (Solari, 2017, p. 13). In principle, the idea is that liberty is best guaranteed through private property in land and markets. Today, many religious principles inspire alternative theorizing about land. Anne Haila (2018) sought to systematically develop this alternative theorizing. Yet as religion increasingly became secularized and economics became religion, markets also became the new vehicle for driving the notion of private property.

The Spread of Private Property: Markets

Unlike religion, the role of markets in spreading private landed property is more recent. The twentieth century, in particular, witnessed the dominance of pro-private property ideas. The cloak of conditions both of morality and of conscience to "help" the poor from religion was stripped off in this era. During this time, Garrett Hardin, whose influential paper "the tragedy of the commons," published in 1968, claimed that common-pool resources tend to be mismanaged because people are individualistic, selfish, and profit oriented. As mentioned in chapter 1, from this perspective, there is a tragedy if there is no individuation and no capitalist markets in land. For Hardin (1968), common property is an aberration and its ills are intensified by population growth. While he considered privatizing the commons may be unjust, he argued that "the alternative of the commons is too horrifying to contemplate. Injustice is preferable to total ruin" (1968, p. 1247). Land tenure systems in Africa were commonly believed to fit this open range system (see, for example, World Bank, 1975, 2003). Being customary is equated with being open range, being undefined, being waste, being frontier, being "no person's land," and being incapable of exchange (Deininger, 2003; Norberg, 2005; World Bank, 1975). These views inspired expensive and extensive land reform programs have been carried out in Africa (Gilbert, 2012).

In *The Other Path* (1989), Peruvian economist Hernando de Soto continues to popularize this pro-private property idea. Focused mainly on ensuring its spread in the countries in the so-called Third World, de Soto argued that the informal economy "other" is *the* path to progress, the only condition being that greater and wider private property had to be created in the land of its residents. So, by using the other path – making land, especially in the informal economy, private property – society would make tremendous progress. The book was widely praised (see, for example, Marquez, 1990), perhaps leading de Soto to make his even bolder claims in the twenty-first century.

In *The Mystery of Capital: Why Capitalism Triumphs in the West and Fails Everywhere Else* (2000), de Soto put the case for individuation in the commons even more forcefully. As is well known, de Soto was inspired by Milton Friedman in 1979 (de Soto, 2004). As crusaders typically do in checking up on their new converts periodically, Milton Friedman sought to make de Soto a total convert when, five years after their meeting, he sent him *The Tyranny of the Status Quo* as a book present. Preaching more private markets and individuation, de Soto was "impressed" by this book, which he subsequently put into practice (de Soto, 2004, pp. 1–2). So successful was de Soto that he won the Milton Friedman Prize in 2004. Elites whose de Soto's ideas have influenced include former US president Clinton, whose endorsement of de Soto's advocacy for private property essentially styles it as the best antidote to the problems of poor societies (Obeng-Odoom, 2013c).

This view has received strong support from the World Bank. Its economists (e.g., Klaus Deininger) and partners such as the International Monetary Fund (IMF) have supported it, too. So, the twenty-first century has been an era in which the commons in the Global South, especially those in Africa and Latin America, have been subjected to the pro-private property doctrine. Advocates imply that the systems are backward and are in need of transformation to a more formal, Western system of land tenure relations. As recently detailed elsewhere (Obeng-Odoom, 2012b, 2020), non-Western countries have been promised considerable prosperity if they shed their system of land use and ownership and embrace a Western – read "individual" – version of securing property rights.

So, the commons in Latin America have been persistently subjected to IMF and World Bank economic doctrines. A notable case is the privatization of the oil commons in Mexico. According to Laguna (2004, pp. 2036–2037), in 1917, the Mexican state took the view that oil, being the commons of the Mexican people, had to be managed by a nationally owned entity with the strong participation of workers. Consequently, the government enshrined this commons view as article 27 in the Mexican Constitution. Thus, when oil exploration started in 1938, The Petróleos Mexicanos (Pemex) was formed to perform this national role. However, over the years, its commons identity has been consistently nibbled away by neoliberal policies, sometimes imposed by the IMF and the World Bank.

A few examples highlight this experience. In 1986, private operators within the nationally monopolized petrochemical industry gained the free hand to obtain inputs on the market if Pemex could not supply them. Next, between 1987 and 1993, over 71,000 workers were laid off and the salaries of remaining workers were reduced by 50 per cent. Then, with the signing of the North American Free Trade Agreement in 1992, the private sector started to enjoy greater access to the commons, including drilling marine wells. In 1995, the Mexican government signed the Agreement on the Oil Income Scheme as part of the Guarantee Agreement with the US Treasury Department that entailed the mortgaging of Pemex's income and, essentially, revealing Mexico's strategic information for a loan facility (Laguna, 2004, pp. 2036–2037).

However, the oil commons has recently been undermined. The country's congress and states, in the second week of December 2013, passed the Energy Reform Bill, which quickly received presidential assent to become law, gazetted on December 20, 2013. The implications of the fundamental changes in the Constitution are not yet fully understood, but one thing is clear: it dissipates and decimates the commons. Eljuri and Johnston (2014) note that by amending articles 25, 27, and 28 of the Mexican Constitution, the new law now makes it possible for substantial private sector participation in the upstream, midstream, and downstream sectors of the oil and gas industry. Notably, it abolishes workers' involvement in the management of the commons, so the Petroleum Workers' Union no longer holds any seats on the Pemex board of directors.

The removal of the common from the commons is also echoed in the marginalization of public opinions on whether to change the Mexican Constitution. A survey conducted by Centro de Investigación y Docencia Económicas before the changes took place showed that 65 per cent of Mexicans were against the decision to enclose the oil commons, but this public opinion was ignored (Estevez, 2013). As with the example of privatizing farmlands in Africa, the argument is framed around efficiency: the commons are inefficient, and their enclosure leads to greater efficiency, in the words of advocates of the enclosure of the Mexican oil commons. The story of the privatization of the commons is clearly complex. The role of money, religion, and markets cannot be overlooked. Together with self-interest and coercion to privatize land, we can think of many other ways in which the commons became commodities. Existing commons continue to be commodified in new ways. Ultimately, the smorgasbord of the "property mind" can entail diverse processes and ways of making private property the "norm" (Haila, 2017). Beyond God and mammon discussed in this chapter, there are other more insidious ways of doing so. The media and the military are two examples (Obeng-Odoom, 2020). These stories also show in what ways the idea of private property spread around the world. This contribution, then, problematizes both the natural story by the Conventional Wisdom and the fable of the enclosure used by the Western Left Consensus.

Privatizing the Commons: Competing Analytical Paradigms

The analytical debates about the commons can further illustrate the point. Paradigmatically, the terms of the original debates about the commons and the creation of property therein were invented by the Greek philosophers more than two thousand years ago. Three issues were recurrent, namely, whether (1) private property is natural or conventional,

(2) common property leads to better use of resources, and (3) private property is more suitable to a happier society and the nature of humans. The debates between the so-called natural rights school and the conventional school have been largely polarized.

Conventional School

As its name implies, the conventional school argued that property in land was conventional, not natural, naturalized rather than primitive. For this school, sometimes regarded as the "socialist group" in Schlatter's (1951) book, the source of society's ills can be traced to the enclosure of the commons. It forcefully argued that common property is more consistent with the primitive nature of humans and the equality of society than is private property. In turn, advocates posited collectivization and a return to traditional form as a panacea to the ills of capitalism. The argument went like this: the common form was the norm, it was negated into a private form, and hence it needed to be negated back to the norm. This is the historically famous Hegel-Marxian idea of "negation of the negation" (Schlatter, 1951, p. 264).

What is also clear is that there is much variation in what we call "political economy of the commons and the common." Take the approaches of Marx, Hegel, and Polanyi. Marx and Hegel were the closest in thought and approach, but even then, Marx's approach is said to be materialist, while Hegel is classified as idealist. Both are dialectical and historical; both use *Aufhebung*, that is, "transformation and incorporation" (Judis, 2013, p. 77); and both subscribed to the approach of "negation of the negation" (Schlatter, 1951, p. 264). Yet, Marx is a historical materialist, not an idealist (Marx & Engels, 1888; Marx, 1867/1990). A fuller account of Marx's political economy of land has been given by Don Munro (2013) in *Journal of Australian Political Economy*. The Hegelians prioritized a three-part approach to enquiry, namely, thesis, antithesis, and synthesis (Schlatter, 1951). However, their more distinctive approach was the emphasis on ideas and history as merely change or episodes. Marx was Hegelian, too, at the beginning that is, but he, like Engels, broke away to found and use a new approach of enquiry which became known as historical materialism. The approach to property was more of an attempt to study how they came to be, and so context and institutions were crucial as were systems and drivers of change. The focus was on the conflictual nature of change and continuity.

Labelled *historismus* and led by scholars like Friedrich Karl von Savigny, this approach was revolutionary, as it set the case of the privatistic school on the back foot, while greatly enhancing the position of the commons advocates (Schlatter, 1951). Still, the historical research did prove that in some jurisdictions, private property in land co-existed with commons, so the main victory for the commons was that it was feasible, although most eventually became privatized. That conclusion was seized upon by Herbert Spencer, a well-known advocate of laissez-faire ideas (Schlatter, 1951). He argued that the tendency to privatization was proof that it was *the* natural way to development, a view supported in contemporary times by the World Bank and the IMF. Spencer came to this conclusion after much prevarication. According to Schlatter, starting in his 1850 *Social Statics* (as cited in Schlatter, 1951), Spencer favoured the nationalization of all common property. Next, in his 1884 book, *Man Versus the State*, he parted company with his own earlier view and started strongly advocating privatistic ownership (as cited in Schlatter, 1951).

Then, in 1891, Spencer wrote *Justice* in which he argued that landowners deserved their rent (as cited in Schlatter, 1951). Overall, his argument was that laissez-faire had a solid place in the commons and their evolution to private property system (Schlatter, 1951). However, additional research, including by anthropologists who followed the historians, revealed that "development" followed such diverse patterns that the Spenserian argument cannot be accurate on the face of the historical record. Its continuing propagation by the apparatchiks of neoliberalism in poorer countries where there is evidence that the commons are more natural can only be a case of "conceptual bias," (Elahi & Stilwell, 2013) or, simply, a deification of private property. Polanyi differs from Marx and Hegel because he places much greater emphasis on the uniqueness of land and rent in the process of creating a surplus.

In this commons system, the lower classes are all co-owners or part owners of the commons, and there is no discrimination along the status of humans: all humans are free and, hence, freely and equally partake in the use of the commons. Plato, Socrates, Cicero, and Seneca all asserted aspects of this view. Within this school, some advocates such as Thomas Rutherford tried to combine elements in the debates and argued instead that some property is natural and, hence, ought to be held in common, while property created by labour ought to be individuated. However, the defining feature of the school, the work of Plato, was the more general sense of community and the commons.

Natural Rights School

In contrast, the natural rights school argues that the institution of property is natural and that common property leads to a dissipation of resources. Private property, on the other hand, is a better way to use resources. It is fairer, as people keep what they produce. In turn, advocates argued that creating a private property system is the best way to organize the commons. A variation of this view is that the state can confiscate common property but then parcel it out over time to make it private property. The natural rights school was argued, notably, by Aristotle who contended that private property is justifiable, among other reasons, because it reduces conflict over who owned what and which belonged to whom (O'Boyle & Welch, 2016).

Followers of Aristotle, notably Thomas Aquinas, also added other reasons for privileging private over common property. They were the encouragement of hard work, the discouragement of sloth, and the expansion of peace and security (O'Boyle & Welch, 2016). Many others, such as Jeremy Bentham, and later the Roman lawyers, canonized these views into Roman law. The tendency was to argue that private property in the commons was primitive, natural, or completely developed in the original state of humans. Some advocates, such as the Roman jurists Hermogenianus and Gaius, argued that even if private property were not natural, it would still be good for society to invent it as it is more consistent with human nature.

The position is aptly summarized by Schlatter (1951, p. 9): "Opponents of private property are foolhardy dreamers." As noted by E. J. O'Boyle and P. J. Welch (2016), these ideas percolated and catalyzed the founding of "personalist economics," which, while quite different from neoclassical economics because of its strong spiritual and moral emphases (e.g., advocacy of generosity and benevolence), holds the view – central in mainstream economics – that "the human person is the basic unit of economic decision-making and economic analysis" (p. 13).

In these debates, the approach of the natural rights school was rationalist, positivist, ahistorical, and class blind from the start. Adam Smith's book, *An Inquiry into the Wealth of Nations* (1776/2007), may have done better, but it avoided in-depth analysis of class and history. It articulated the doctrine of *homo economicus* – the idea of the self-interested or selfish individual. It is problematic, though, to equate Smith's oeuvre to what has come to be called "methodological individualism," mainly because Smith's *Wealth of Nations* must be understood in the wider context of his *Theory of Moral Sentiments* (1759/2005), which he had published earlier. Accordingly, political economists (e.g., Elahi & Stilwell, 2013; Garnett Jr., 2019; O'Boyle & Welch, 2016) note that Smith was not entirely sanguine about the emerging capitalist system, which, he argued, would not produce equality and would, in fact, be exploitative and problematic.

Yet, as shown by Schlatter (1951), the Smithian labour theory of value contains elements of the Lockean theory of property. Smith hints at but avoids the radical conclusions about class conflict arising from the distribution of rent. Ricardo and others (e.g., Mill) saw these conclusions too, but avoided them. The conclusion that anyone who profited from the work of others was an exploiter was not asserted, although it is the logical end of the labour theory of value as seized upon by the Ricardian socialists. Rev. Samuel Newman, Rev. Francis Wayland, and Henry Carey were some of the leading classical economists who held onto the theory of value but avoided its radical implications. Eventually, the labour theory of value was abandoned.

The "abandonment" of the classical school was not because of its inherent weaknesses per se. The "danger" that logical conclusions from its analyses would unsettle the dominant classes in society made this school quite dangerous. As noted by Howard Sherman (1993), replacing classical economics with neoclassical economics, another school that focused more on the tendency towards harmony (equilibrium) festooned with the appearance of science through the use of tools such as mathematics, became increasingly appealing to the power groups who stood to gain the most from the turnaround. In turn, the work in this direction began. The "marginalist revolution," pioneered by Jevons, Menger, and Walras and greatly boosted by the work of Alfred Marshall, led to the insularization and mathematization of economic science (Elahi & Stilwell, 2013). This twist paved the way for a completely new way of doing economics in the 1870s with significant advances in the 1950s (Stilwell, 2012a, p. 61). The role of Hayek, Friedman, Stigler, and others in the Mont Pellerin Society and the University of Chicago Economics Department is detailed elsewhere (see Jones, 2010; Mirowski, 1988a, 1988b) as is the impact of influential studies by Paul Samuelson among others (Elahi & Stilwell, 2013, pp. 31–33). The key point here is that, after its emergence in the nineteenth century, neoclassical economics by the twentieth century had become mainstream, with new institutional economics developing much later in the late twentieth century and early twenty-first century.

The emergence of neoclassical economics from classical economics could be called a "scientific revolution" in Kuhnian terms. However, the rise of new institutional economics was more of a "scientific advance" in the Popperian sense. This emergence represented a critical juncture of mainstream thinking, not a critical disjuncture or departure from its core values. New institutional economics, then, was a reformist advance, not a revolutionary change. Even though it emphasized exchange over choice (the key focus of the neoclassicals), as Peter Boettke and his colleagues (2012) have shown, with its continuing commitment to efficiency, for example, new institutional economics was, in effect, neoclassical economics in disguise.

One widely used explanation is derived from the "open access exploitation thesis," which predicts that without private and formal property rights, a resources boom will inevitably lead to widespread socioeconomic problems in a resource-rich economy (Barbier, 2005, pp. 122–140; de Soto, 2011). Public institutions and the state are assumed to lack the expertise to manage natural resources or, alternatively, are assumed to be so entrenched in a culture of corruption that they are not in a position to effectively manage such resources. This view, widely regarded in neoclassical economics as *the* property rights approach to natural resources, was popularized by Armen Alchian and Harold Demsetz (see, for example, Alchian & Demsetz, 1973; Demsetz, 1967; Demsetz, 2002). It was well regarded at the time of its emergence for breaking away from the established body of work on the theory of production and exchange that draws largely on a negatively sloping demand curve, with only punctuated concerns about abnormal demand conditions (Pejovich, 1972).

Sometimes styled as the "new property rights school," the dependence of new institutional economics on neoclassical economics methodology, values of competition and individuation, and policies for laissez-faire society is exemplified in the work of Douglas North, James Buchannan, and Ronald Coase (Dugger, 1980; Fine, 2010a). The new property rights school, and its neoclassical economics school masters such as behavioural economics, qualify but still emphasize *homo economicus* in "the property rights paradigm," to use the title of two of its most respected analysts, Armen Alchian and Harold Demsetz (1973). The effort is most eloquently highlighted in the book *Oil Is Not a Curse: Ownership Structure and Institutions in Soviet Successor States* (Luong & Weinthal, 2010). By introducing questions of ownership structures, how they evolve over time, and institutional drivers and consequences, the authors suggest that they have made a giant departure from the neoclassical economics framework.

Concerns about institutions, the structure of property rights, and how they emerge and evolve have always been a feature of the orthodox property rights tradition. Indeed, Armen Alchian and Harold Demsetz (1973) observed in their seminal paper on the property rights approach that the three defining questions of the approach are "(1) What is the structure of property rights in a society at some point of time? (2) What consequences for social interaction flow from a particular structure of property rights? and (3) How has this property right structure come into being?" (p. 17). In answering these questions, restrictive neoclassical economics assumptions are retained (see, for example, de Soto, 2000, 2011), including the tendency towards equilibrium in a market society, the rational individual, the profit-maximizing individual, and the presence of sufficient information in all societies at all times. Markets were extolled as *the* best medium of social organization, while normative arguments were made in favour of pricing, privatizing, and marketizing society, environment, the relationship between humans and natural resources, and the environment in which they co-exist. (For a critical discussion of orthodox economics methodology and approaches, see, for example, Butler et al., 2009, pp. 105–117; Keen, 2003; Spies-Butcher et al., 2012; Stilwell, 2012a, 2019a).

Neoclassical and new institutional economists come to conclusions that defend and extend private property in land. They predict doom for any ownership structure that resembles, even faintly, communal or customary property rights (Luong and Weinthal, 2010). Yet as we have seen in our analysis of Richard Schlatter's *Private Property: The History of an Idea* (1951), this shared position in the commons argument predates neoclassical economics. Over the years, however, many economists defended and canonized it into a formal approach to analysis in economic science. A detailed history of how this approach became dominant in the economics discipline has been offered by Harold Demsetz (2002), one of the leading figures in the property rights school in economic science. I will not repeat it here.

The point is that this property rights approach accepts the status quo without systematically asking how we came to have one class as landlord, another as capitalist, and a third as worker. Culminating in what I call the Conventional Wisdom in this book, it makes no distinction between land and the rest of the factors of production, as the principle of substitutability is king (for elucidation of this critique, see Gaffney, 2008; Stilwell & Jordan, 2004a, 2004b).

It is this line of thinking – whether of the neoclassical or new institutional variety – that has mostly been rewarded with the Nobel Prize in Economic Sciences. In their analysis of the Nobel Prize in Economic Science, Peter Boettke and his colleagues (2012) show that, since 1969 when the first prize was given, usually, the winner is either a neoclassical economist or a new institutional economist. As Avner Offer and Gabriel Söderberg (2016) have shown, this distribution of the prize is neither accidental nor innocent. Rather, it reflects the underlying class structure of society, legitimizes the political preference of dominant classes, and extols private property in land over the common property in land.

Conclusion

The historical context analysed in this chapter brings out important lessons. First, in contrast to the prevailing literature that emphasizes eighteenth-century debates or even twentieth-century debates of the commons and property, this chapter has shown that as early as the third century, the debates had started. Not only were they heated and rancorous, but they also overlapped with the emergence of money, and they developed in lands as distant from Africa as Greece and Rome, although they have eventually strongly influenced the idea of land in Africa.

Second, the debates about privatizing the commons is both historic and global, but they have been shaped by the uneven geographies of the world. While economists have been part of the debate, the terms have been framed not only by them but also by philosophers and religious and legal scholars. Third, the Conventional Wisdom has increasingly become more powerful and more widely accepted as *the* scientific standard in the twenty-first century, influenced by particular political-economic interests. So, the *transition* to the Conventional Wisdom of today could not have been brought about by apolitical scientific advances, but rather politicized "scientific revolution" as per Kuhn. It is the constellation of interests that has combined to propel and sustain the Conventional Wisdom whose rise to power roughly overlaps with the decline of classical economics, the rise of neoclassical economics, and the recent power of new institutional economics. As Ben Fine (2010b) has noted, this Conventional Wisdom has expanded its scope by incorporating the arguments of opponents and including areas often overlooked by neoclassical economics. However, this embrace of "others" is, in effect, a kiss of death because it has been done to advance the mainstream, imperialize the "others," and defend the image of orthodoxy.

Consequently, political economists have provided a large literature that is critical of this school of economics. The tendency (see, for example, Haila, 2016; Milonakis & Meramveliotakis, 2013) has been to focus their searching studies on Ronald Coase, Douglas North, Armen Alchian, and Harold Demsetz, the pioneers of new institutional economics. However, the work of the Bloomington School of New Institutionalism (see Aligica & Tarko, 2012, p. 237), led by Elinor Ostrom, has received far less systematic study. The next chapter turns to a study of the Bloomington School of the Commons, the crème de la crème of the Conventional Wisdom today.

Chapter Three

Rethinking the Commons

Introduction

In his historic 2015 encyclical letter, "Care for Our Common Home," the Pope of the Catholic Church forcefully makes the case to reconsider the meaning, prospects, and future of the commons. "The growing problem of marine waste and the protection of the open seas," writes the pontiff, "represent particular challenges." He continues, "What is needed, in effect, is an agreement on systems of governance for the whole range of so-called 'global commons'" (Francis, 2015, p. 28).

The commons has always been a major window through which political economists have viewed the capitalist world system. Indeed, some would begin their political-economic analysis with the (in)famous "enclosure of the commons" question, drawing attention to or inspiration from the accounts in Karl Marx's *Capital* (1867/1990), Karl Polanyi's *The Great Transformation* (1944/2001), or Maurice Dobb's *Studies in the Development of Capitalism* (1946). However, modern analysis of the vitality of the commons begins, and often ends, with the work of Elinor Ostrom, who won the Nobel Prize in Economic Sciences for her work on the commons, or, to be precise, "for her analysis of economic governance, especially the commons" ("Elinor Ostrom: Facts," 2009, para. 5). Although credited with offering the global and transdisciplinary analysis that decisively showed the fallacies in Hardin's (1968) "the tragedy of the commons" (see, for example, Harvey, 2011), Ostrom's work was better known in political science than in political economy. Indeed, her own analysis of research in her area and the sources of ideas on which she drew in her lifetime are mostly from political science and mainstream economics, as her work "Traditions and Trends in the Study of the Commons" (van Laerhoven & Ostrom, 2007), in the first issue of *International Journal of the Commons* shows. In turn, many political economists asked "Elinor who?" when her name was mentioned as the Nobel Prize in Economic Sciences winner in 2009 (Stilwell, 2012a, p. 45).

After she won, Ostrom became the centre of much attention, with both orthodox and heterodox writers seeking to appropriate her work. Awards have been instituted in her name. For instance, *International Journal of the Commons* gives the Ostrom Memorial Award to the best papers published in the journal, while the *Journal of Institutional Economics* gives the Elinor Ostrom Prize. In *Environmental Markets: A Property Rights Approach*, T. L. Anderson and G. D. Libecap (2014), two well-known mainstream environmental economists who advocate the institutionalization of private property rights in nature, note that "our understanding of how common property institutions constrain over-harvest or over-extraction owes much to the work of ... Elinor Ostrom" (p. 94). For heterodox thinkers, Ostrom's appeal appears to lie in the fact that she wrote supportively of the commons, a holy grail of progressive scholars. In turn, much academic scholarship in heterodox economics cites Ostrom's work favourably. The influential Cambridge economist Ha-Joon Chang (2011), for example, places her among heterodox writers on institutions:

> It is news to me that Ostrom has ever belonged to the orthodox institutionalist circle. She is a political scientist who has mostly propagated her ideas through books – that low-grade activity that orthodox economists tend to despise. Most of her journal publications are, naturally, in political science journals and most of the economics journals she has published in were heterodox ones, that is, until she got the Nobel Prize. The reactions shown by young USA-based economists in the American Economic Association's job search website, www.econjobrumors.com, in the days after the announcement for her Nobel Prize are a very good, if absolutely shocking, testament to the contempt in which she is held by most mainstream economists. (pp. 609–610)

In a special issue on Ostrom published in the *Journal of Institutional Economics* (Hodgson, 2013, p. 383), the editor of the journal placed Ostrom's contribution in political economy, arguing that her methodology, work, values, and thinking make her more heterodox. Also, in a special issue on common property that appeared in the *Review of Radical Political Economics*, Christopher Gunn (2015) approvingly used Ostrom's principles to organize his thoughts about the commons. Similarly, the leading German heterodox economist, Wolfram Elsner, with his team of heterodox economists, endorse Ostrom's work in their impressive textbook *The Microeconomics of Complex Economies* (Elsner et al., 2014). Although there are critics from the heterodox schools of economics, they are few (see, for example, Fine, 2010b). In a sense, this (mis)classification of Ostrom's work is as much as indication of the breadth of heterodoxy and its contested merging with orthodoxy at one extreme (Hodgson, 2014; Obeng-Odoom & Bromley, 2020).

For rhetorical purposes, the embrace of Ostrom's work may seem effective. However, this chapter shows why political economists ought to be cautious, indeed wary, of uncritical endorsement of Ostrom's work. In spite of claims of overlaps between Ostrom's methodology and that of heterodox institutional political economists (Hodgson, 2013), Ostrom described herself as a "new institutional economist," not an institutional political economist. Although she cited the pioneer of institutional political economy J. R. Commons in her book *Governing the Commons* (1990), the key influence on her was the work of the new institutional economists, and the audiences she most enthusiastically engaged were liberal-mainstream rather than radical-heterodox thinkers. Indeed, Ostrom gave the Hayek lecture in 2012, organized by the right-wing Institute of Economic Affairs in London (Ostrom, 2012b). "Ms Elinor Ostrom," wrote well-known Mexican activist Gustavo Esteva (2014, p. i147), "was a very sweet and dedicated lady. But she was pretty ignorant. She lacked historical perspective and empirical information about her theme, the commons." If some empirical work affirms Ostrom's principles, other work (especially Bolognesi & Nahrath, 2020; Gerber et al., 2009) contradicts them, showing that, in fact, the more the principles are applied, the greater the tendency for conflict to arise.

Critical institutionalists such as Frances Cleaver and Jessica de Koning have also stressed the limitations of Ostrom's mainstream orientation, pointing to its narrow foundations in rational choice analysis. For such critics, a more hybrid, dynamic, and pluralist framework of "institutional bricolage" can help us to better understand the complexity of the commons (see, for example, Cleaver, 2002, 2012; de Koning, 2011, 2014; Cleaver & de Koning, 2015; Sarker & Blomquist, 2019).

This chapter extends this critical perspective. It accepts that invoking the work of Ostrom has convinced mainstream economists that collective governance of the commons is possible without the attendant challenges of the tragedy of the commons and free-rider problems. However, it argues that a more systematic appraisal of Ostrom's work shows that it is hardly an avatar of the commons and society. Although questions of ecological and social justice are central to addressing the planetary crises (Bromley, 1991, 2019; Hiedanpää & Bromley, 2016), Ostrom's work contains no systematically developed concept of justice.

She frequently mentions rights in her work, but their conception is extremely limited, often tied to the notion of joint, rather than equal, rights. For Ostrom, the notion of the commons is socially separatist, partially economic, and environmental but not socioecological. Ostrom's analysis of the commons is commendably historical, but it is hardly historiographical. Conceptually, she confuses *common property resources* with *common-pool resources*. The latter was the focus of her PhD on investigating groundwater problems in California. Tragically, she applied this analysis to the wider issue of common property resources, a slippage facilitated by the fact that both are "CPR" questions (Obeng-Odoom & Bromley, 2020). Riddled with these conceptual and concrete problems, her proposed "collective action" to save the commons actually provides fertile grounds for the real threats to the commons.

With even its most visible framework crippled by such deep problems and with environmental economics evidently limited, Amartya Sen (2015), a critical voice within mainstream development economics, recently acknowledged the helplessness of the Conventional Wisdom:

> Environmental analysis is seriously hampered by not having anything like an adequately broad normative framework, involving ethics as well as science that could serve as the basis of debates and discussions on policy recommendations. Despite the ubiquity and reach of the environmental dangers, a general normative framework for the evaluation of these dangers has yet to evolve. (p. 8)

A strikingly different and more holistic approach to the commons is offered by Henry George, who posits the commons as *the* most important path to social, economic, and ecological sustainability. Unlike Ostrom, who studied the commons as a "scientist" who wanted to show "scientifically" that there is another good that is neither private, public, nor club-based, George studied the commons to understand and remove injustice at the roots. In turn, his approach is more critical and certainly more relevant today in showing that another world is possible, even if both historical and contemporary examples of the commons and how they function(ed) suggest that George's work, too, requires significant changes to update its framing of the meanings, prospects, and future of the commons.

This analysis implies that the encyclopedic definition of the commons – "common property resources usually refer to ubiquitous or fugitive resources which appear to be in the public or even global domain without any clear structure of ownership or control" (Dragun, 2001, p. 118) – ought to be called into question. Both the orthodox view that private property rights systems are needed to save the commons (Anderson & Libecap, 2014, p. 95) and the heterodox economics argument that "ironically, the solution to 'open access resource problems' appears to be the establishment of more formal property rights structures closer to common property rights than pure private rights" (Dragun, 2001, p. 118) are grossly oversimplified. The commons are historically specific and contextually defined in meaning and prospects, hardly ubiquitous and unchanging. Similarly, the commons are neither saved by private property rights nor are they defined by "formal property rights structures."

Threats to successful commons have come from external aggression, not from some lack of consensus and from over-exploitation by greedy societies or the self-interest of users *in vacuo*. Either way, however, the threat to them has always been force, either militarily, through deceit, or by economic pressure on the communards. In historical examples in which communards have mishandled the commons, the cause was usually the pressure to make good on loans advanced to communards under usurious terms (Ciriacy-Wantrup & Bishop, 1975; Cobb, 2016; Obeng-Odoom, 2021).

Capitalism and imperialism continue to be the most enduring threats to the future of the commons. To propose that capitalist institutions rescue the commons is, therefore, to issue their death warrant and expedite their extinction. The rest of the chapter is divided into three sections. The next section appraises the work of Elinor Ostrom. It is followed by an analysis of the strengths and weaknesses of Henry George's work on the commons. The last section examines the future of the commons based on the analysis of both Ostrom and George.

The Mainstream Foundations of Elinor Ostrom's Principles on the Commons

It is important to revisit the philosophical basis of Elinor Ostrom's work because, although Ostrom herself is well known, the deeper meaning of her work is widely misunderstood (Sarker & Blomquist, 2019). Ostrom's idea of the commons comes from James Buchanan's notion of "club goods," as distinct from Paul Samuelson's idea of "pure" private and "pure" public goods. According to Buchanan's economic theory of clubs, "The interesting cases are those goods and services, the consumption of which involves some 'publicness,' where the optimal sharing group is more than one person or family but smaller than an infinitely large number" (1965, p. 2). The "theory of clubs" is, in effect, "a theory of cooperative membership" (1965, p. 1).

Ostrom accepts Buchanan's conception (see also Tarko, 2012, 2017), making only minor modifications to terminology (from "club" to "toll" goods; from "rivalry consumption" to "subtractability of use"), methodology (switching from a zero-sum game in terms of subtractability to a positive sum game where subtractability is on a continuum), and conception (replacing "club" with "common-pool resources") (see Ostrom, 2009a, p. 412). In her Nobel Prize Lecture (Ostrom, 2009a), Ostrom recounted how and why she came by the concept of "common-pool resources." It is telling that it is not driven by political-economic concerns but by a desire to challenge analytical categories. Her major achievement, then, is demonstrating that beyond the institutions of "market" and "state," there can be a third institution: the community. Her argument is that goods are not simply "public" or "private" goods, they can also be "common-pool resources" with the characteristics of both public and private goods, although these features need not necessarily be attributable to private or public goods (Ostrom, 2009a, 2012b).

These common-pool resources, Ostrom stressed, could escape the well-known mainstream economics concerns: free-rider (the usual concern about public goods) and tragedy of the commons (the typical criticisms of common-pool resources). Those resources that, in fact, escaped these problems provided Ostrom the blueprint for her famous institutional analysis and development (IAD) principles.

The aim of the framework was to guide further studies by Ostrom's team: The Workshop in Political Theory and Policy Analysis. For other authors, the IAD framework "is intended to contain the most general set of variables that an institutional analyst may want to use to examine a diversity of institutional settings including human interactions within markets, private firms, families, community organizations, legislatures, and government agencies. It provides a metatheoretical language to enable scholars to discuss any particular theory or to compare theories" (Ostrom, 2009a, p. 414). Aside from developing a framework, Ostrom made important policy suggestions, too. Instead of designing institutions to coerce individuals to act in certain ways, she argued in her Nobel Lecture that it is much better to support the creation of those institutions that unleash the best in human potential. In this sense, she favoured the design of institutions to help bring out the best in individuals in terms of learning, trust building, cooperation, and innovation. The support for such institutions, Ostrom consistently argued (see Tarko, 2012, 2017, for biographical essays, but also Ostrom, 2009b, 2012b), must be polycentric and multi-scalar. It would seem that, for Ostrom, as for many other welfare economists, the basis for thinking about the allocation of goods and governance is *consumer sovereignty*. Individual preferences are always sacrosanct in her work on the commons, sharply contrasting with the work of political economists such as Richard Musgrave, who developed the concept of "merit goods": "'Admittedly difficult to define and dangerous to entertain, communal concerns,' wrote Richard Musgrave, 'have been part of the scene from Plato on, and my concept of merit goods … was to provide a limited opening for their role'" (Musgrave, 1997, p. 30, as cited in Desmarais-Tremblay, 2019, p. 230). Ostrom's consumer sovereignty approach is more consistent with the Conventional Wisdom. It is, as a result, more widely accepted by the Establishment. Elaborating her ideas is, therefore, important.

If any one book can capture the essence of Ostrom's work, it is *Governing the Commons: The Evolution of Institutions for Collective Action* (Ostrom, 1990). The book challenges apocalyptic views that eventual destruction is the way of all commons. It argues that the commons are governable, notably at the subnational level. An important book, it goes beyond the existing polarized views on how such governance might be exercised either through the market or through the state. Instead, it offers a model of governance based on individuals working jointly together and for a common end. For these individuals to succeed, the book advocates a set of principles through which they can govern the commons.

Ostrom's conclusions in her book are based on a simple, but highly important, method of data collection: systematization of existing research. She does not plunge into fieldwork straight away. Her work is based first on careful scrutiny of studies on the questions she seeks to understand. These case studies are particularly important. They cover different contexts, and they were carried out by people from diverse disciplinary backgrounds. After then choosing the most detailed existing research, Ostrom and her team then probed deeper into selected cases by doing additional fieldwork to better understand and determine (1) successes, (2) failed commons, (3) how success stories evolve, and (4) the keys for successful governance of the commons. This intriguing approach has much to teach social scientists, who often rush to do fieldwork when data exist. The lesson is that it is further analysis and systematics that are required in such cases. It is from this approach that we derive Ostrom's concept of the commons and her principles about how they are governed.

The commons in Ostrom's work are referred to as "common-pool resources (CPRs)." In her words, "The term 'common-pool resource' refers to a natural or man-made resource system that is sufficiently large as to make it costly (but not impossible) to exclude potential beneficiaries from obtaining benefits from its use" (Ostrom, 1990, p. 30). For Ostrom, the commons have two parts: the resource system as a whole

Table 3.1. Ostrom's Design Principles for Successful Governance of the Commons

Number	Principles
1.	*Clearly defined boundaries:* Individuals or households who have rights to withdraw resource units from the CPR must be clearly defined, as must the boundaries of the CPR itself.
2.	*Congruence between appropriation and provision rules and local conditions:* Appropriation rules restricting time, place, technology, and/or quantity of resource units are related to local conditions and to provision rules requiring labor, material, and/or money.
3.	*Collective-choice arrangements:* Most individuals affected by the operational rules can participate in modifying the operational rules.
4.	*Monitoring:* Monitors, who actively audit CPR conditions and appropriator behavior, are accountable to the appropriators or are the appropriators.
5.	*Graduated sanctions:* Appropriators who violate operational rules are likely to be assessed graduated sanctions (depending on the seriousness and context of the offense) by other appropriators, by officials accountable to these appropriators, or by both.
6.	*Conflict-resolution mechanisms:* Appropriators and their officials have rapid access to low-cost local arenas to resolve conflicts among appropriators or between appropriators and officials.
7.	*Minimal recognition of rights to organize:* The rights of appropriators to devise their own institutions are not challenged by external governmental authorities.
For CPRs that form part of larger systems	
8.	*Nested enterprises:* Appropriation, provision, monitoring, enforcement, conflict resolution, and governance activities are organized in multiple layers of nested enterprises.

Source: Ostrom, 1990, p. 90.

and the resource units. The stock refers to the structure, such as oceans, areas for grazing, and lakes, or the outwardly human-made resources such as a bridges, while the resource units connote the rewards from the resource systems. The people using the commons can be appropriators (those using the resource units), providers (those seeking to expand the system), or producers (those responsible for calling the commons by actually producing it in the beginning). The relationship among all of them is one of "joint use" (Ostrom, 1990, p. 31). The common principles of success for all the CPRs studies constitute the famous Ostrom IAD framework, described in Table 3.1.

This story of the commons and how they are governed is spread out in the six chapters of Ostrom's book. The research problem and how it has been dealt with are the focus of chapter 1. CPR features and methodology are discussed in chapter 2. Successful empirical cases are analysed in chapter 3 of her book. The historical analysis of change to successful models is in chapter 4, while chapter 5 is devoted to a discussion of failed cases. The theory of how to address the problems identified in chapter 1 is developed in chapter 6.

Ostrom wrote and spoke clearly and widely about her subject and convictions, which helped to make her contribution well known. Between 1990, when *Governing the Commons* was published, and 2012, when Ostrom died, she wrote extensively to document her work on the subject (see, e.g., Ostrom, 2007, 2009a, 2009b; Ostrom & Basurto, 2011). While she held on to her core arguments and principles, Ostrom did make some minor changes to her principles. For instance, after studying the outcome of research based on her principles published in *Ecology and Society* in 2010, she expanded her principle on clarity of boundaries to clarity of boundaries *and* clarity on users' boundaries (Ostrom, 2012b).

Others have helped to propagate her work too. Frischmann (2013, p. 387) has summarized Ostrom's "two enduring lessons" as "a substantive lesson that involves embracing complexity and context, and a methodological lesson that involves embracing a framework-driven approach to systematic, evolutionary learning through various interdisciplinary methodologies, theories, and empirical approaches," while for Gunn (2015, p. 3), "The lessons derived from her work are that management of the commons must be by rules established by those who use it, and that there is a collective private property alternative to individual private property or government regulation that has worked over time." What is rarely realized, however, is that Ostrom's idea of the commons was without justice, society, or economy. Being a separatist concept, Ostrom's commons has little, if any, role in social, economic, and ecological transformation. The emphasis on internal threats to the commons without a similarly extensive analysis of external threats to the commons further leaves Ostrom's commons without justice, society, or economy.

Commons without Justice, Society, or Economy

Nowhere in Ostrom's *Governing the Commons* is there a theory of justice, and the index of the book contains neither the word nor a semblance of it. Ostrom's Nobel Prize Lecture (2009a) contains only two mentions of the word *justice* (Ostrom, 2009a, pp. 431, 435) and two mentions of *sustainable/sustainability* (Ostrom, 2009a, pp. 435, 436). Although central to the planetary socioecological crises (see, for example, Bromley, 1991; Bromley, 2019; Hiedanpää & Bromley, 2016; Obeng-Odoom & Bromley, 2020), Ostrom does not consider matters of just change, social, or ecological justice, let alone their intersections. The place of the commons in the broader society-economy-ecology milieu is hardly established by Ostrom. If Ostrom has no theory of growth, she has even less to say about change, the distribution of wealth, and the rich-poor gap that belies much global poverty. In turn, Ostrom's concept of the commons fares only fairly well in terms of how it helps us to understand society and how the economy works.

Part of the reason why this approach tends to be taken is that Ostrom preferred respectable, "scientific" research in the mainstream and so used a superior mainstream model to counter existing ones. In the words of one admirer: "Ostrom was a scientist. Her response to concerns about model-induced myopia was to do the scientific work of systematically studying actual resource systems and governance institutions" (Frischmann, 2013, p. 392). A second reason is that for Ostrom, the commons are a mere analytical category, but there is a most fundamental reason: Ostrom's mainstream politics. It plays out most clearly in her strategy of saving the commons to sell them.

Saving the Commons to Sell Them

Ostrom's contribution to ecology is much better than her contribution to justice, economy, and society, but it is still limited. She was preoccupied with avoiding free-rider problems and showing how decentralized communities can govern their natural resources. The principles of the commons end up seeking to reduce "transaction costs," much like Coasian principles. When Ostrom turns to climate change and the environment (Ostrom, 2008, 2009b), her main interest is in making sure that everyone plays a role in reducing emissions. As she explains herself in another context, "Instead of a single best design that would have to cope with the wide variety of problems faced in different localities, a polycentric theory generates core principles that can be used in the design of effective local institutions when used by informed and interested citizens and public officials" (Ostrom, 2010d, p. 112).

On the day of her death, she published an opinion piece, "Green from the Grassroots." She argued in this work that from subnational to national and then to international units, all should be involved in seeking to cut emissions. Against the current effort about international emissions standards, Ostrom (2012a) argued that "grassroots" made up of layers beneath the international should all play their role to mitigate the climate crisis and adapt to a changing climate. This practical concern is important, of course, for no one action can reduce climate change. Hence, Ostrom's case for polycentricity is well considered.

What is curious is that her argument assumes that contributions to climate change are similar among different interest groups, institutions, and entities. She recognizes that richer countries are more polluting (Ostrom, 2008), but demurs from making a similar charge against capitalist firms. Yet much empirical evidence compiled and analysed in "Challenging Climate Change," a special issue of the *Journal of Australian Political Economy* (see for a summary Goodman & Rosewarne, 2011), as well as a special issue on "Environmental Impacts of Transnational Corporations in the Global South" in *Research in Political Economy* (see, for a summary, Cooney & Freslon, 2018), shows that such firms and corporations, particularly those in Western societies, are the most polluting and carbon emitting; justice will require that they make the most sacrifices. It is revealing that, on the one hand, Ostrom's reviews of the evidence of systems of governance of CPRs (see Ostrom, 2008, 2009b) are very favourable towards establishing private property rights and markets in the commons through mechanisms such as quota management systems, total allowable catch, and individual transferable quota. Yet, on the other hand, Ostrom's reviews are mostly negative towards state-based intervention or even community-wide ownership of the commons.

In celebrating the work of Ostrom, Pennington (2013, p. 449) notes that Ostrom's work "lacks a robust account of when, if ever, top-down governance arrangements are to be preferred." Indeed, in spite of the attempt to avoid the commitment to any one institutional position (Pennington, 2012; Tarko, 2012, 2017), there is a systematic but subtle attempt to delegitimize the state in the work of Ostrom, what Pennington (2012, p. 31) euphemistically calls "a presumption against central planning." The reasons for this presumption are the same ones used by mainstream economists against the use of the visible hand of the state: namely, a supposed inability of the state to correctly know what is best for society (or the utilitarian argument that individuals always know what is in their best interest and will act to maximize it; see Cobb, 2000, pp. 7–14, for exposition), inefficiency in the management of resources, and distortion of incentives. Much like what Friedrich Hayek (1945) argued in his *Road to Serfdom*, the tendency in Ostrom's work is to suggest that the state should be a helper of the market, for example, through protecting private property rights of CPR owners (Pennington, 2012, pp. 31–38), not through supporting the community.

In Ostrom's words, "instead of presuming that one can design an optimal system in advance and then make it work, we must think about ways to analyse the structure of CPRs, how these change over time, and adopt a multi-level, experimental approach rather than a top-down approach to the design of effective institutions" (Ostrom, 2008, p. 29). Much like the push for "urban governance" or "self-help" programs (Obeng-Odoom, 2013d, 2020), such arguments amount, in practice, to restricting the state to cede more and more space to markets, and making the state pro-market. It is a case of "in markets we trust," as a special issue of *Ecological Economics* on market-based instruments for environmental sustainability shows (for a summary, see Gómez-Baggethun & Muradian, 2015).

In fairness to Ostrom, she is more cautious in her analysis of the commons and their ecological contributions. Indeed, she argued against a "panacea trap" (2012b, p. 69) – the view that there is only one solution. Instead, she advocated a "polycentric" panacea. The goal of Ostrom's work is "the development of a broader theory of institutional arrangements related to the effective governance and management of CPRs" (Ostrom, 1990, p. xiv). The central question for Ostrom is how to develop self-government for a CPR. In her own words, "The central question ... is how a group of principals who are in an interdependent situation can organize and govern themselves to obtain continuing joint benefits when all face temptations to free-ride, shirk, or otherwise act opportunistically" (Ostrom, 1990, p. 29). The concept of "property rights" in the commons, from this perspective, is simply the rights of individuals over CPRs. The rights discussed are not equal but joint rights.

In this sense, the rights behave like a joint account to which all account holders have a joint claim, and one person cannot access the account without the consent of the others, as Henry George and many Georgists have consistently argued. For instance, Richard Giles (2017) notes that equal rights are direct rights, while joint rights are derivative and contingent. The two types of right might, but do not necessarily, lead to the same outcome. To give another example, land rights and the right to land might bring about similar outcomes by accident, but the superior "right to land" is not the same as the idea of land rights. What Ostrom seeks is the minor land rights, say to land in a gated community.

Whether capitalism will allow these enclaves to exist forever is a difficult question for Ostrom and her followers. Ostrom has a theory of "collective action," but it is individual based. The basis of the discussion of rights is the individual, and the emphasis is on self-help or organized rules and duties. These institutions govern the commons to avoid "over use" (Ostrom, 1990, p. 32) through what is frequently called "joint use" (e.g., Ostrom, 1990, p. 31). The basis of the theorizing is the same "rational individual" as in mainstream economics, albeit expanded to include the possibility that the rational individual is not merely opportunistic but can cooperate with others to enhance their cost-benefit utility maximization. In her own words: "I use a very broad conception of rational action, rather than a narrowly defined conception ... Individuals selecting strategies jointly produce outcomes in an external world that impinge on future expectations concerning the benefits and costs of actions" (Ostrom, 1990, p. 37).

This classic individual-based analysis or what Ostrom calls "the internal world of individual choice" (1990, p. 37) requires further comment. In Ostrom's conception, whether the commons can be saved depends on four factors: expected costs, expected benefits, internal norms, and discount rates. An individual will reflect the prevailing attitude of other individuals, perhaps shaped by the world[1] or by other factors, and, together with rational calculations on the expected cost and benefit of cooperation using internal discount rates, decide whether to cooperate. Every time, therefore, the individual makes decisions about independence, interdependence, and collective action. Ostrom's key argument is that individuals do cooperate without the need of the firm (market) or the state, as it known in conventional economics, but through self-help (see Ostrom, 1990, pp. 38–43). Curiously, though, the emphasis is on how this self-help generates some equilibrium: a combination of rational choice analysis and some ideas about institutionalism (Ostrom, 1990, p. 43). Ostrom (1990) said it best:

> Given the similarity between many CPR problems and the problems of providing small-scale collective goods, the findings from this volume should contribute to an understanding of the factors that can enhance or detract from the capabilities of individuals to organize collective action related to providing local public goods. All efforts to organize collective action, whether by an external rule, an entrepreneur, or a set of principals who wish to gain collective benefits, must address a common set of problems. These have to do with coping with free-riding, solving commitment

---

1 For Ostrom (1990, p. 206), this "world" influence is suggested to be "situational variables," namely, living near CPRs, the experiences of appropriators working together on other issues, and available information about opportunities outside the resource system.

problems, arranging for the supply of new institutions, and monitoring individual compliance with sets of rules. (pp. 27–28)

In short, *Governing the Commons* is "a study that focuses on how individuals avoid free-riding, achieve high levels of commitment, arrange for new institutions, and monitor conformity to a set of rules in CPR environments [and] should contribute to an understanding of how individuals address these crucial problems in some other settings as well" (Ostrom, 1990, pp. 27–28).

Of course, this emphasis, too, is important. The problem is that in seeking to achieve this end, the book exhibits a limited conception of time and history. There is a recount of history as a series of events, notably judicial decisions to affirm or reject rights related to common or individual property (chapter 4), but these do not drive Ostrom's analysis. Indeed, the cases in chapter 4 of her book are not systematically related to the successful cases in chapter 3, where Ostrom declares, "For the cases that I discuss in chapter 3, I do not know what the structures of the situations were like before some appropriators in the mists of time began to experiment with various rules to allocate resource units and provisioning responsibilities" (1990, p. 56). In fact, *temporal* in Ostrom's work is equated to *geographical*. Ostrom does not address the history of the "success" or "failed" cases she analyses. Yet, driven by the quest to disprove the "tragedy of the commons" (Hardin, 1968), she turns her attention mainly to the internal forces of change. From Ostrom, then, increasing contemporary land and water grabs that leave many hungry, dispossessed, and landless (Elhardary & Obeng-Odoom, 2012; Obeng-Odoom, 2013b, 2014a, 2015c, 2021) are not a problem as long as they did not emanate from internal over-exploitation. As the IAD shows, Ostrom's focus is mainly internal threats. What about society-wide threats? What role do the commons play in the economy, ecology, and society? Ostrom offers limited answers on these issues. One theorist who took these issues seriously and addressed them systematically and extensively was Henry George.

Henry George and the Fundamental Principle of the Commons: Equal Access to Land

Unlike Ostrom, Henry George was a scholar, a public intellectual, and an activist who was, at least once, imprisoned for his radical ideas. World famous and a pillar of a global movement, George was, perhaps, second in importance only to the likes of Mark Twain, the American novelist, and Thomas Edison, the renowned scientist (Cobb, 2015). Henry George has been called "America's greatest early economist" (Bryson, 2011; Cleveland, 2012, p. 509), and his analysis of the commons stands in sharp contrast to Ostrom's.

For Henry George, the commons represent nature – indeed, the equal rights to own, access, use, and control nature, as distinct from "the results of labor [, which] should belong to him who has labored" (1891, p. 15). At the heart of Henry George's conception of the commons is the notion of justice, what he called "the justice of common ownership" (1891, p. 2). This concept of justice is grounded in the idea of *natural rights*. In chapter 10 of *Social Problems* (George, 1966), George identifies two sets of rights, the denial of which lead to social problems. One set relates to the commons; the other, to the private individual. By recognizing both, George (1891, p. 24) argues that humans have both a "social nature" and an "individual nature" – features that interlock to make the human being a "land animal," that is, one reliant on common land. The role of the state, according to George, is to (1) ensure that the results of labour are due to labour, (2) ensure that land/nature is common for all, and (3) take steps to prevent the destruction of the commons or their conflation with the private.

For George, the commons is land. However, as Richard Giles (2017, pp. 49–50) correctly points out, because some lands have been privatized, George (1886/1991) talks of making such land also common property "where no one could claim the exclusive use of any particular piece" (p. 279). George's conception of the commons has sometimes been called "nature's storehouse" for all of us – as Pullen (2014) reminds us in his book *Nature's Gifts*. These are our commons for three important reasons. First, they are given by nature; that is, they are not the private property of anyone. Second, they are common because the rights exercised over them are fundamentally rooted in the notion of equal rights, as distinct from joint rights. This means that humans have inalienable rights to natural resources and inalienable rights to resources created by communal labour and effort (see also George, 1892/1981). These rights are limited only to the extent that they infringe on another person's rights to access and control of such resources. From this perspective, Georgists (e.g., Subere-Albawy, 2015) insist that land is the primary commons. *Land* here refers to mines and minerals under the physical land, waters over and under land, and the atmosphere, the environment, or the air we breathe. Georgists have also argued that publicly created goods, such as collectively produced parks, should also be regarded as commons in the sense that no one individual can lay claim to them while excluding others from their use. Third, land is or should be made commons in the sense that it is the commonwealth from which *social, spatial,* and *ecological* problems should be resolved. In his famous lecture "Moses," Henry George (1884) sets out in detail the various ways in which land can be used to address common problems. According to him, land should be used to create

> a commonwealth based upon the individual – a commonwealth whose ideal it was that every [wo]man should sit under [her or] his own vine and fig tree, with none to vex [her or] him or make [her or] him afraid ; a commonwealth in which none should be condemned to ceaseless toil; in which, for even the bond slave, there should be hope; in which, for even the beast of burden, there should be rest. A commonwealth in which, in the absence of deep poverty, the manly virtues that spring from personal independence should harden into a national character – a commonwealth in which the family affections might knit their tendrils around each member, binding with links stronger than steel the various parts into the living whole. (p. 9)

The commons in the Georgist sense are inextricably linked to commodities. It is a "bitter irony," George argues in *Social Problems* (1883/1966), "to place a man where all the land is appropriated as the property of other people and to tell him that he is a free man, at liberty to work for himself and to enjoy his own earnings" (p. 99). Both commodities and the commons must be guaranteed as *natural rights* to ensure a good society in which there is "progress without poverty" (George, 1879/2006), the removal of social problems, and the prevention of the great social theft. In George, this reads as: "There are only three ways by which any individual can get wealth – by work, by gift or by theft. And, clearly, the reason why the workers get so little is that the beggars and thieves get so much" (George, 1883/1966, p. 84). As beggars can hardly be wealthy, George clarifies, "When a man gets wealth he does not produce, he necessarily gets it at the expense of those who produce it" (1883/1966, p. 84), including beggars and workers, leaving capitalists and landlords as the great exploiters.

In the Georgist conception of the commons, therefore, there is a strong emphasis on how the commons cement society to the economy and ecology. The root cause of social change, according to George, is contingent on how the commons are handled. If there is a denial of equal rights to the commons through either forced or contrived enclosure for private, colonial, or imperial uses, there is adverse social change in which much wealth coexists with poverty (Obeng-Odoom, 2015a).

This injustice is compounded if there is a denial of the rights to private property in labour through extracting a surplus from the fruits of labour, individually but also collectively. Social problems such as hunger, insecurity, and begging are all the result of this denial of common rights in the soil and private rights in the products of labour (George, 1883/1966).

As recently suggested by Peirce (2015), George also saw the commons as contributing to or being an obstacle to economic growth, depending on how they are used. Focusing, in particular, on cities and trade, and how agglomeration, specialization, and the application of technology generate economies, George showed that when the urban commons are increased in size, the economy, as a flow process, expands, too. When more land is brought from hoarding into the commons, output is likely to increase, especially if the full rights of labour to its produce are also enabled. George (1892/1981) accepted exclusive *possession* of parts of the commons to the extent that such possession would help to encourage and to secure the exclusive *ownership* of the products of *labour*. In his words, "While the right of ownership that justly attaches to things produced by labor cannot attach to land, there may attach to land a right of possession. Private possession of the land on which labor is thus expended is needed to secure the right of property in the products of labour." He continues, "This right of private possession in things created by God is however very different from the right of private ownership in things produced by labor … The purpose of the one, the exclusive possession of land, is merely to secure the other, the exclusive ownership of the produce of labor; and it can never rightfully be carried so far as to impair or deny this" (George, 1891, p. 2).

George stood for a common that powered growth, what he called "progress" in publications such as *Progress and Poverty* (1879/2006). To ensure growth and prosperity, George (1891) noted:

> We have no fear of capital, regarding it as the natural handmaiden of labor; we look on interest in itself as natural and just; we would set no limit to accumulation, nor impose on the rich any burden that is not equally placed on the poor; we see no evil in competition, but deem unrestricted competition to be as necessary to the health of the industrial and social organism as the free circulation of the blood is to the health of the bodily organism – to be the agency whereby the fullest cooperation is to be secured. (p. 25)

George's focus, however, was on what Cobb (2015, p. 462) – echoing the voices of many Georgists – has called "cooperative competition." This is competition with a socioecological face and for socioecological purposes. George's theory of growth works through unleashing innovation and entrepreneurship and curbing unproductive rent-seeking practices (Baumol, 2004; Birkeland, 2020). In this process, progress without poverty is unlocked and socioecological growth unchained.

Contrariwise, if the commons are enclosed, such economies become concentrated in the hands of a few, which, in turn, creates poverty amid plenty. The precise mechanisms by which economic concentration arises in the Georgist analysis are many. Speculation, a direct corollary of turning the commons into private property, for example, is directly responsible for creating a land price bubble. Such escalating land prices tend to (1) swallow wages, which decline as a proportion of the general surplus, regardless of improvements in labour-enhancing technology or new labour skills, which might be expected to help labour to be more productive and be rewarded, (2) increase housing prices, and (3) price out essential services.

These three collectively tend to generate inequality and poverty. They also stimulate low production, as labour becomes disinterested in production, among other reasons, because even improvements in labour-saving technology do not necessarily bring labour more wages. In turn, labour may become incapable of purchasing what is produced, setting in motion economic crises that can culminate in increased crime. Inequality and poverty can also increase. In turn, governments can become more corrupt, as the landed rich use their wealth to control all aspects of the state, including the government (George, 1883/1966, pp. 14–15). The consequences of this web of socio-economic problems are many. Social levels of happiness can be choked off. Economic fragility can be unchained. Climate can change. As Gaffney (2015) has shown, these processes were very much at play in the last global economic downturn, with its wide-ranging effects for society, economy, and ecology.

Much of the "economics of Henry George" (Bryson, 2011; Cleveland, 2012) tends to overlook George's theory of sustainable development, but there has been important work on this aspect too (Daly et al., 1994; Obeng-Odoom, 2016c; Stilwell & Jordan, 2004a, 2004b). This scholarship links George's notion of the commons to sustainability. Only greater commoning can save the planet in this sense. Private interest in the commons tends to make speculation attractive, which tends to increase land rent. In this process, goods and services for the rich, who are able to pay for central locations, push services that are essential for the poor out to the peripheries because providers of such services are not able to pay the increasing rents. Georgists (e.g., Subere-Albawy, 2015) have implied that speculation on land itself increases the distance between where we want to go and where we live and, hence, the need to drive. When speculators hoard land that can be used for the provision of social services, such amenities will have to be located farther and farther away from where people live; this raises the likelihood that driving becomes an imperative. Increasing rent forces people to move much farther from essential services and again raises the likelihood that people will have to drive to get to work and to access basic amenities. At the ecological level, speculation has been established to be detrimental to ecological justice. It reduces space for parks and greens, urban farms, and gardens. It pushes greens apart from where people live, resulting in pollution, which can arise from commuting to such parks or essential services. Private property in the commons also leads to over-exploitation of resources because of competitive private use for profit. In short, without the commons, ecosystems are broken, urban agriculture is despised, people are incapable of paying for space in the city, and urban sprawl is encouraged (e.g., Obeng-Odoom, 2013a, 2016d, 2020; Subere-Albawy, 2015).

The result of implementing George's *uniform* land value tax would be to discourage urban sprawl, which arises because developers or individuals seek development at the outskirts, where there is low or no tax on land development (Stilwell & Jordan, 2004b). Land value taxation can also discourage the proliferation of vacant lots, which arises from speculation. Finally, in *For the Common Good*, Herman Daly and his colleagues (1994, p. 258) demonstrate that to reorient the economy to community, the environment, and a sustainable future, a land value tax is needed. Seeking to socialize privately appropriated socially created rents, such a tax would fall on land values, not on human exertion. Without this site value tax, Herman Daly and his colleagues (1994) argue, urban farmers are likely to oppose pro-environmental zoning because they stand to benefit from selling "their" land to developers. With a Georgist tax, the roots of opposition to ecological zoning would be removed, paving the way for more sustainable cities.

In one sense, Georgist proposals on the environment amount to what has been called "selling the environment in order to save it" (Stilwell, 2011) because they work through the price and selling mechanism. However, there is a small, but important distinction between George's proposals and the mainstream economics approach to selling the environment. George's true remedy is "nothing short of making land common property" (1879/2006, p. xvii), whereas for mainstream economists, the true remedy is privatizing the commons. Figure 3.1 demonstrates this important difference, showing that Henry George points

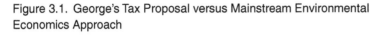

Figure 3.1. George's Tax Proposal versus Mainstream Environmental Economics Approach

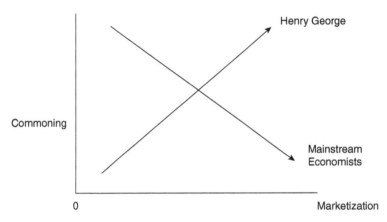

towards greater commoning, while mainstream economists' move further from commoning and deeper into marketization. In the words of Herman Daly and his colleagues (1994, p. 256), George gravitates towards a "stewardship" conception of the commons or land, while mainstream economists seek an "ownership" view of land.

George recognized that the only solution to the crises of humanity had to be the recognition of the equal "rights of man." These rights entailed commons rights and individual rights. George (1891) advocated a two-prong approach: making all land common property *and* removing all taxes on labour:

> We propose – leaving land in the private possession of individuals, with full liberty on their part to give, sell or bequeath it – simply to levy on it for public uses a tax that shall equal the annual value of the land itself, irrespective of the use made of it or the improvements on it. And since this would provide amply for the need of public revenues, we would accompany this tax on land values with the repeal of all taxes now levied on the products and processes of industry – which taxes, since they take from the earnings of labour, we hold to be infringements of the right of property. (p. 3)

Mainstream economists and others ignore the essence of Georgism. They consider, instead, that George was merely proposing efficient public administration, steeped in taxation reform (see, for example, Behrens et al., 2015). As George (1891) himself noted, the reform proposed has both ethical and economic aspects that go hand in hand in a

union, although "the ethical is the more important side ... the benefi-
cent and far-reaching revolution we aim at is too great a thing to be
accomplished by 'intelligent self-interest'" (p. 8). Taxation is but a tool
in the social reform George seeks: to give to all what belongs to all
(land) and give to labour what belongs to labour (the products arising from the exertion of labour). In *Social Problems*, George (1883/1966)
discusses common rights to urban soils and community gardens and
parks as a fundamental solution to urban social problems: "With the
resumption of common rights to the soil, the overcrowded population of the cities would spread, the scattered population of the country
would grow denser" (p. 238). George's plans for making privatized
land also a commons or common property is to levy a tax on land
value and remove all taxes from labour. When this is done, George
(1883/1966) notes:

> In a society where the equality of natural rights is recognized, it is mani-
> fest that there can be no great disparity in fortunes. None except the physi-
> cally incapacitated will be dependent on others; but there can be no very
> rich class, and no very poor class; and, as each generation becomes posses-
> sed of equal natural opportunities, whatever differences in fortune grow
> up in one generation will not tend to perpetuate themselves. In such a
> community, whatever may be its form, the political organization must be
> essentially democratic. (p. 194)

The Common Sins of Henry George and the Georgists?

George is often accused of committing three sins, namely, focusing only
on land tax, endorsing the Lockean theory of common property, and
misunderstanding the nature of communism and Indigenous commons. As these are widespread criticisms, it is important to consider
each of them in turn.

LAND TAX AS THE ONLY REMEDY

Georgist political economy is critical of land tax. It considers land *value*
tax, that is, the rents on land, not physical land itself. That is rather
straightforward to deal with, although it has often been said that Georgism is all about the taxation of rent and that George claimed that all
social problems will disappear with taxation of ground rents (Behrens
et al., 2015; O'Sullivan, 2012, pp. 150–151). In fact, George's argument
is more sophisticated. He understood that there were other forms of
inequality, including racism and sexism, and did not think that land
value tax would end all of these. However, his argument was that any

solution that ignored a comprehensive analysis of the commons and how their privatization undermined equal rights would be limited:

> Let me not be misunderstood. I do not say that in the recognition of the equal and unalienable right of each human being to the natural elements from which life must be supported and wants satisfied, lies the solution of all social problems. I fully recognize the fact that even after we do this, much will remain to do. We might recognize the equal right to land, and yet tyranny and spoliation be continued. But whatever else we do, so long as we fail to recognize the equal right to the elements of nature, nothing will avail to remedy that unnatural inequality in the distribution of wealth which is fraught with so much evil and danger. Reform as we may, until we make this fundamental reform, our material progress but can tend to differentiate our people into the monstrously rich and the frightfully poor. Whatever be the increase of wealth, the masses will still be ground toward the point of bare subsistence – we must still have our great criminal classes, our paupers and our tramps, men and women driven to degradation and desperation from inability to make an honest living (George, 1883/1966, p. 201)

George, then, admits the analysis of, say, trade unions and supports workers, their unions, and activities. As Edward O'Donnell's book *Henry George and the Crisis of Inequality* (2015) shows, George was active in labour politics and activism. The candidate of the Workers' Party, the United Labor Party, he had substantial support from urban poor workers (see, for example, chapter 5, "labor built this republic, labor shall rule it" of O'Donnell's book) and from the Knights of Labor, of which he was a prominent member. Nevertheless, George is clear that the tactics of labour are limited because they do not look at rights in land, only individual rights in labour. Focusing on George's activities rather than his principles of political economy or on his principles without looking at his political activities is part of the reason for the myth that George and Georgists only focus on land tax.

GEORGISM AS LOCKEAN: GEORGE ACCEPTED ASPECTS OF LOCKE'S ANALYSIS

In *The Perplexed Philosopher*, George (1892/1981) notes:

> Locke was not in error. The right of property in things produced by labor – and this is the only true right of property – springs directly from the right of the individual to himself, or as Locke expresses it, from his "property in his own person." It is as clear and has as fully the sanction of equity in

any savage state as in the most elaborate civilization. Labor can, of course, produce nothing without land; but the right to the use of land is a primary individual right, not springing from society, or depending on the consent of society, either expressed or implied, but inhering in the individual, and resulting from his presence in the world. Men must have rights before they can have equal rights. Each man has a right to use the world because he is here and wants to use the world. The equality of this right is merely a limitation arising from the presence of others with like rights. Society, in other words, does not grant, and cannot equitably withhold from any individual, the right to the use of land. That right exists before society and independently of society, belonging at birth to each individual, and ceasing only with his death. Society itself has no original right to the use of land. What right it has with regard to the use of land is simply that which is derived from and is necessary to the determination of the rights of the individuals who compose it. That is to say, the function of society with regard to the use of land only begins where individual rights clash, and is to secure equality between these clashing right of individuals. (p. 33)

However, with reference to Locke's claim about the commons being privatized by mixing labour with natural goods, George approves of Herbert Spencer's discussion of Locke on this point: "The reasoning used in the last chapter to prove that no amount of labor, bestowed by an individual upon a part of the earth's surface, can nullify the title of society to that part might be similarly employed to show that no one can, by the mere act of appropriating to himself any wild unclaimed animal or fruit, supersede the joint claims of other men to it" (George, 1892/1981, p. 23).

George (1892/1981) certainly agreed with Spencer that the application of labour can never be used to justify individual privatization of the commons:

What Locke meant, or at least the expression that will give full and practical form to his idea, is simply this: That the equal right to life involves the equal right to the use of natural materials; that, consequently, any one has a right to the use of such natural opportunities as may not be wanted by anyone else; and that the result of his labor, so expended, does of right become his individual property against all the world. For, where one man wants to use a natural opportunity that no one else wants to use, he has a right to do so, which springs from and is attested by the fact of his existence. This is an absolute, unlimited right, so long and in so far as no one else wants to use the same natural opportunity. Then, but not till then, it becomes limited by the similar rights of others. Thus no question of the right of any one to use any natural opportunity can arise until more than one man wants to use the same natural opportunity. It is only then that any question of this right, any need for the action of society in the adjustment of equal rights to land, can come up. (p. 33)

So, George does not support this Lockean view that appropriation of commons is permitted merely by adding labour to it.

Where George errs is in his broader reading of Locke. This became clear when George challenged Pope Leo XIII's contention "that industry expended on land gives ownership in the land itself, and that the improvement of land creates benefits indistinguishable and inseparable from the land itself" (as cited in George, 1891, p. 14). To this claim, George wrote: "Your contention is not valid. Industry expended on land gives ownership in the fruits of that industry, but not in the land itself" (1891, p. 15). George had a clearly formulated stance on what is, in fact, as much Leo as Lockean – but George erred in not realizing that Pope Leo's position was analogous to Locke's. Locke, a slaver in the slave trade (which George opposed and warned the Pope against), offered a theory of property (common and private) that aimed at justifying property in slavery, justifying the dispossession of Indigenous people, and justifying the private appropriation of common land through some notion of hard work (see Alexander & Peñalver, 2012, pp. 35–56; Denman, 1978, pp. 17–18; Duchrow & Hinkelammert, 2004, pp. 47–67; Pullen, 2013). George stood for a different kind of politics though, so a total reading of George would easily show that his analysis of the commons is not Lockean.

GEORGE AGAINST COMMUNISM AND INDIGENOUS PROPERTY RIGHTS

George considered himself different from the communists – either voluntary or forced communism. Communism aspires to the abolition of *all* private property; Georgism seeks the abolition of only *private property in land*. Communism is voluntary if people willingly live in common. Forced commons entails the use of state power to force people to live in commons. Of the two, George (1891) preferred voluntary communism, calling it "the highest possible state of which men can conceive" (p. 23). George argued that the commons under communism can flourish, but at the time of his writing, he considered the commons to be only artefacts of an earlier era and no longer possible in the modern/his era unless there religious faith were widely diffused among the people: "But we see that communism is only possible where there exists a general and intense religious faith, and we see that such a state can be reached only through a state of justice" (1891, p. 24). Analytically,

George takes the communists to task when they advocate "thoroughgo-ing socialism" because, for George (1891), "it fails to see that oppression does not come from the nature of capital, but from the wrong that robs labor of capital by divorcing it from land, and that creates fictitious cap-ital that is really capitalized monopoly" (p. 25). Communism, George argued, undermines growth, kills individual initiative, and leads to political despotism.

While George is silent about the place of Indigenous property and Indigenous people in his philosophy, his principles clearly suggest that they, as first-comers or autochthons, have no special rights compared to other human beings. George holds that land is for all at all times and for all humans. The story George tells, that of Cain and Abel, is aimed at establishing the point: first-comers can enjoy exclusive rights but when latecomers arrive, they too become equal owners (1891, p. 2). Rights to the use and ownership of land, indeed land, are equal and eternal from the Georgist reading, so Indigenous people have no special con-sideration in a Georgist society. It does not follow, however, that Indig-enous populations would be expelled from the land under a Georgist regime because for George *ownership* of the commons is not needed for social progress. Rather, some rights of *possession* are all that is required to blend progress with biodiversity and the maintenance of Indigenous cultures. Still, the ethical concern about Indigenous peoples' autoch-thonous rights pose a difficult riddle for Georgists. Stilwell and Jor-dan (2004b, p. 12) have proposed that under a Georgist notion of the commons, Indigenous people might be exempted from tax or made the recipients of substantial shares of the additional revenues generated from tax either in direct cash, social services, or a combination of these – as one form of acknowledgment of their special connection to the land.

As with all targeted proposals, however, it will have to overcome two significant administrative problems: the problem of inclusion and the problem of exclusion (Segal, 2011, p. 478). The former arises when the benefit erroneously leaks to people who are not Indigenous, while the second problem of exclusion arises when Indigenous peo-ple are mistakenly left out of the scheme. As research on the "idea of Indigenous people" (Béteille, 1998) and "the social identities of young Indigenous people" (Jang, 2015) shows, identifying who is an Indig-enous person can be extremely difficult, often triggering debates about whether it is better to rely on self-identification or external, bureau-cratic stock-taking and the implications of either mechanism. In the literature, the way to avoid exclusion problems is to make a benefit universal, while inclusion problems can be addressed by giving no one the benefit (Segal, 2011, p. 478).

Clifford Cobb (2016, p. 287) has more recently suggested as a possible solution that Georgists can rework their unit of analysis to focus more on cultures. From this perspective, the autochthons can welcome second comers to share their land, but the latecomers will have to recognize that their own rights are inferior to those of Indigenous people. Indeed, that approach will chastise individualistic cultures in favour of more collective cultures. The challenge, as Cobb points out, is that this could be reverse racism. Yet, in principle, because it extols collective ways of life in relation to the land – which George endorsed – the issue of reverse racism is merely academic.

A concrete proposal for the twenty-first century – echoing this approach – has been given by Stilwell and Jordan (2004b). They suggest holding consultations with Indigenous leaders and communities in a consultative democratic process in determining Indigeneity and how to share rents. While this route might be time-consuming and cannot resolve the issue of reverse racism, it is, perhaps, the best Georgism can do about the thorny issue of first- and second- comer rights, which are complex in practice.

Take *Obshchina* in pre-revolutionary Russia – a peasant commons organized around at least seven principles (see details in Barnett, 2004; Ely, 1916; Grant, 1976; Kimball, 1973). First, land was held in common, with the peasant having equal, not joint, use rights in the commons. The community held the freehold interest. Second, the common land was periodically redistributed to keep the commons in balance. Third, the *Obshchina* was dynamic, not static, so commoners sought to expand the principles and practices beyond the commons, such that between the eighteenth and nineteenth centuries, the *Obshchina* expanded greatly. Indeed, *Obshchina* was internalized when Russian radicals sought to draw out principles of the commons for international socialism. Fourth, the leadership of the commons was grassroots but hierarchical. The hierarchy, however, entailed some form of elections. Fifth, commoners had fiscal obligations. That is, they paid dues and taxes to help in the social development of the commons. Sixth, the commons had a philosophy of egalitarianism, which was anti-capitalist in content, nature, and history. Seventh, the *Obshchina* must always be discussed with the *Mir*, the political organization that promoted equity in the community of peasants, because often there were overlapping and reinforcing practices. Eighth, the biggest threat to the commons came from the expansion of capitalism, not individual "tragedy of the commons."

While a revolutionary state helped to support and even expand the *Obshchina*, it took the state working with capital to destroy the *Obshchina*. The passage of the *ukas* in 1906, based on individual private property in land supported by state credit, state lands, and other statist efforts to transform the *Obshchina* into a statist, individual-based system, led eventually to its collapse. Such commons entail other aspects of sociality such as judicial, executive, and administrative activities for which taxation is necessary but not sufficient as a substitute, so it opens up a grey area in Georgism.

One example of a commons arrangement which offered training in legislative, executive, and judiciary activities was "the Mark," most commonly found in Germany but also elsewhere such as in Sweden (Engels, 1892/1928). In these places, people organized into villages, hundreds of which collectively formed a *gau*. The Mark itself had to be made up of at least six villages. Common property in Indigenous societies served the society excellently: it provided a good locomotive for the Indigenous economy and offered an excellent ecological support system

In this socio-economic organization, land was held in common and households worked different aspects determined by casting lots among the people. These parcels of land were actively redistributed, for example, every 3, 6, or 12 years, depending on need to ensure that there was no land concentration. Also, the land was tilled using the principle of crop rotation. Each allocated land had to practise three-field farming: within the year, one lot had to be partitioned into three, one each given to the cultivation of winter crops, summer crops, and fallow season – a system that worked on a rotational basis: a part of the triad was at one point in fallow cultivation but in another time in winter or summer crop cultivation. After distribution, land that is not under cultivation is held in reserve as a village commons or, as Engels (1892/1928) put it, "the uncultivated land, forest and pasture land, is still a common possession for common use" (p. 3). Also held in common were resources found under the surface of the soil.

Generally, the economic organization of such Indigenous systems was good in the sense that they supported the welfare of their people. Research (Ciriacy-Wantrup & Bishop, 1975; Hill, 1961; Obeng-Odoom, 2013b, 2013e, 2015d, 2021) reveals four important insights about such economies in Africa. First, detailed local knowledge of the environment enabled the society to devise the most appropriate tools for work. Second, the absence of speculation meant that inefficiencies associated with private property systems were eliminated. Third, the presence of active redistribution systems discouraged excessive accumulation, which, in turn, kept economic growth at levels lower than in advanced capitalist societies – but not at stagnating levels. Indeed, the levels of poverty and hunger found in modern capitalist societies were hardly a feature of pre-capitalist Indigenous societies. Finally, population pressure on the commons was contained not through Malthusian natural famines but rather through customary rules that regulated marriage, childbirth, and lactation. Outside the family, fission, or the breakaway of small groups to start new communities, also served to contain pressure on the commons. As we will see, these societies are not all extinct: they have evolved and many were malleable even at their origin.

The Mark, for instance, was political, adaptable, and dynamic. Indeed, "the time of sowing and of reaping should not be left to the individual but be fixed for all the community or by custom" (Engels, 1892/1928, p. 6). In some cases, and over certain periods of time, even the produce from the soil was shared in common. There were no written laws, but the customs having been jointly agreed to were binding. As there were no kings or royal nobility, there was "equal share in the legislation, administration and jurisdiction within the mark" (Engels, 1892/1928, p. 6). Meetings were held in the open and frequently to decide on the running of the Mark, including breaches of customs and rules. Decisions were made by "the aggregate of the members present" (Engels, 1892/1928, p. 7). According to Engels (1892/1928), "just as the members of the community originally had equal shares in the soil and equal rights of usage, so they had also an equal share in the legislation, administration and jurisdiction within the mark" (p. 6). Self-governance in the Mark, then, connoted true democracy.

George might have been correct about the nature of "prosperity" in Indigenous societies, but he erred in arguing that such economies were stagnant. Also erroneous was George's attribution of dictatorship to communism/socialism and his equation of communism to statism. George's (1891) comment about communism/socialism is revealing: "Socialism in all its phases looks on the evils of our civilization as springing from the inadequacy or in harmony of natural relations, which must be artificially organized or improved. It its idea there devolves on the state the necessity of intelligently organizing the industrial relations of men, the construction, as it were, of a great machine whose complicated parts shall properly work together under the direction of human intelligence" (p. 25). This organization George (1891) called "the organization of men into industrial armies, the direction and control of all production and exchange by governmental or semi-governmental bureaus," which would invariably lead to "Egyptian despotism" (p. 25), a subject he discussed extensively in his lecture "Moses" (George, 1884). George knew of many such communal organizations and included examples of them – in Campagna, Italy, and the Gaelic tenure in Scotland (see George, 1891, pp. 13–14) – in his open letter to Pope Leo XIII, but his emphasis was on how they had been decimated by private property rights in land. In turn, George was inattentive to, indeed misrepresented, the political organization in *res communis*.

George suggested that any viable alternatives to his proposal were only historical. He analysed actually existing commons. In particular, he considered the Californian Goldfields from which he determined three critical features of the commons (Giles, 2017, pp. 51–52). The first was equal access to land. The second was an acknowledgment of labour's enormous contribution to the creation of wealth and, hence, the need to guarantee that contribution. The third was a bottom-up approach to devising, revising, and enforcing the arrangements for governing the commons *within* a wider framework that reserves a place for other institutions (e.g., states and markets) to recognize and uphold the first two principles. In doing so, it prevents speculation and monopoly or keeps away what Tania Murray Li (2014) calls the "land assembly," or in the words of Richard Giles (2017), those lawyers, surveyors, and banks who work to undermine the commons by seeking to commodify or privatize it.

Indigenous Conceptualizations of the Commons

Many more commons abounded then, and abound in our own time, too. The commons among Indigenous peoples in Australia are often cited as examples, but there are others in the Melanesian region (Anderson, 2011; Boydell, 2010; Haila, 2011, 2018) and in Africa (Obeng-Odoom, 2012a, 2012b, 2021), where the commons are formulated around autochthony or common property in which nature is the giver of land, often on a first-come-first-served basis. However, the idea of autochthony freezes an adaptable system that allows nurture to change the course of nature. In *Land, Mobility, and Belonging in West Africa,* Carola Lentz (2013) shows that among the northerners of Ghana, the earth priest (*Tin daana*), a member of the first-comers who keeps an "earth god" to symbolize the rights of Indigenous people over and above all other latecomers, sometimes "sold" the "earth god" to latecomers. In this process, the "natural" autochthons, the "true" daughters and sons of the soil, through negotiation, pass on the head rights to the commons to newcomers, notably the Dagaras in the East of the Black Volta areas of Northern Ghana. From this perspective, the head rights to the commons are not only the inalienable property of the Indigenous people (Sisala first-comers) but can be transferred under certain conditions.

Similarly, in Brikama in the Gambia, through long service of labour and socialization, latecomers can become first-comers. In *Land, Labour and Entrustment: West African Female Farmers and the Politics of Difference*, Pamela Kea (2010, 2013) analyses how the commons in the Gambia in

West Africa, particularly the activities of female Indigenous landowners, operate. She shows how through a lengthy period of entrustment, latecomers can also become landowners. Entrustment or the *karafoo* refers to a relationship between first-comer land rights holders (hosts) and latecomer (stranger) Gambian female landless labourers. In this arrangement, first-comers who have land but lack labourers enter into a social relationship with latecomers who have labour but no land. After working on their farms for a while, the latecomers get plots of land on their own. It seems that some strangers in Brikama in the Gambia can "filter up" to become owners – through long service and socialization (Kea, 2010, pp. 44, 51, 2013).

As Kea (2010) notes, "Agrarian clientelist relations, although based on unequal access to the means of production, ultimately facilitate a relationship of land and labour sharing between groups of female farmers, allowing recent migrants to be incorporated into larger support networks" (p. 12). On this basis, she concludes that not all patron-client relationships are bad or unproductive. Both patrons (female hosts) and clients (female strangers) are productive and benefit from entrustment (Kea, 2010, pp. 153–163, 167–186, 2013). In this sense, labour and wider socialization could be pathways to becoming members of the land commons, but the labour is deployed entirely for the development of the first-comer rather than merely dispossessing the first-comer, developing an area of first-comer land for their selfish-interest, and then claiming that because they have been on the land for some time, they are co-owners. In turn, these latecomers become first-comers by operation of social forces and gifts (Obeng-Odoom, 2021).

Can Modern Societies Claim to Have Commons?

Some suggest that the notion of the commons end with historical commons, Indigenous commons, and commons in Africa, the Pacific, and a few other places or with history. However, Boydell and Searle (2014) write about "the contemporary" commons, as contrasted with what Andro Linklater (2013, p. 40) refers to as a "primitive commune." Indeed, many political economists (e.g., Harvey, 2011; Newman, 2015) would simply consider public places, spaces, and facilities as "commons" and public activities in urban areas, "urban commons." In this nomenclature, the commons equate with the "public," human-made social practices like "culture," and activities done in common, people working in common, and such (Pithouse, 2014). While such ideas have been helpful, for instance, in grounding claims about shrinking "commons" (e.g., shrinking public housing, transport, and health care), they lack analytical precision.

Table 3.2. Types of Goods and Their Characteristics

		Excludability of Free-Riders	
		Easy	Difficult
Consumption Rivalry	*Large*	1. Private goods	3. Commons
Consumption Subtractability	*Small*	2. Club goods	4. Public goods

Source: Adapted from Tarko, 2012, p. 58.

As Table 3.2 shows, the four types of goods in new institutional economics differ substantially in terms of the effect of consumption on what remains for others (consumption rivalry/subtractability) and whether free-riders can be excluded.

Based on these criteria, the commons are quite distinct from club goods. So, when he is writing about "the future of the commons," David Harvey (2011) claims that "the rich these days have the habit of sealing themselves off in gated communities with which an exclusionary commons gets defined" (p. 103), what he means, analytically at least, is club goods. In their contribution to *Housing Studies*, Tony Manzi and Bill Smith-Bowers (2005), identified – correctly – "Gated Communities as Club Goods"; not the commons. What is called the commons is also distinct from public goods. Even in terms of scale, the state can – and in a Georgist understanding, should – support the commons but its domain of influence tends to be with the provision and maintenance of public goods, while communities tend to be responsible for the commons. The commons can have public attributes, of course, and in some instances, the commons are also public land, such as urban spaces where much informal economic activities occur (see, for example, Obeng-Odoom, 2011a), but to equate the commons with the public can raise significant analytical tensions.

Much of this conflation arises from stressing "property" in the idea of "common property" to the essential neglect of the political economy of the commons. From this fascination with "property," especially "property regimes," Dan Bromley (1992) famously claimed that "common property is corporate group property ... Corporate group property regimes are not incompatible with private, individual use of one or another segment of the resources held under common property" (pp. 11, 12). Bromley is careful to put the situation in context. Yet it is important to more explicitly stress the radical politics of the commons. The radical political economist Christopher Gunn (2015) has offered what he considers to be a more apt description of the commons:

Commons are what we share. They are creations of either nature or society that belong to us equally, and that come with need for a commitment to maintain them for future generations. In their collective aspects they are the antithesis of individually privatized property, which has been at the heart of capitalism for centuries. (p. 1)

Based on the analysis thus far, a few questions could be raised about Gunn's otherwise succinct definition. The suggestion that commons belong to us equally dangerously plays with the notion of "joint right," rather than the idea of equal rights in Henry George and traditional notions of land (Giles, 2017, pp. 55–56). The use of the conjunction "or" in his second sentence where none is needed is another issue. With the "or" struck out, the definition of the commons becomes: commons are the creations of nature to which society has made modifications. This definition upholds the Georgist interpretation but corrects its simplistic assumption of "raw nature." Other implications of the Georgist conception of the commons, such as ensuring that "arrangements for use are set and protected by the members" (Giles, 2017, p. 53), can usefully be emphasized. In these ways, the revised conception of the commons can embrace much discussed commons that are nature's gifts nurtured by society and humans.

Consider the experiences of Alaska, Singapore, and New Mexico. Widerquist and Howard (2012a, 2012b) have shown how Alaska's oil commons is held together by applying Georgist principles. The evidence is clearly that patterning the collection and use of oil rents after Georgist principles has brought much growth and progress in Alaska without the yoke of poverty. Anne Haila's recent work (2016) in Singapore also shows that the public use of publicly captured land rent in Singapore is what has transformed this important city-state. As Gunn's (2015) work shows, *Acequias* are also another example, although of a slightly different genre. Numbering around 900, *Acequias* are water commons in New Mexico and Colorado in the United States, communities that collaborate to share water resources. Such communities have been around for centuries and they continue to manage the commons successfully in economic, social, and ecological terms.

The commons have provided a much-needed support for their agrarian livelihoods with some irregular or no wage employment. *Acequias* are also ecological. They are based on engineering mechanism that relies less on machines and more on traditional embankments and gravity to distribute water among members. Individual members maintain their share of *Acequias*, but they participate in annual communal labour to repair and maintain the *Acequias.* In another sense, the farmers share seeds in a communal spirit. Decisions about running the commons are made collectively through an elected commission of *Acequias* water right holders, although operational matters are handled by an elected, sometimes appointed, mayor (mayordomo) for and on behalf of those in the commons called *parciantes*, a major driver of ecological sustainability in the commons in nature. These commons are run and supported by a culture that is both pre and anti-capitalist.

Community members contribute to an *Acequias* community's ditch funds, into which some states also make some contributions. The state involvement makes *Acequias* quite a "formal" commons, but they are not public or state commons. Neither do they owe their survival to state formal laws. Instead, Hispanic culture that tries to be consciously against individual privatization has helped to ward off, as Gunn (2015) discovered in his fieldwork, persistent pressure on *parciantes* to "sell" their water rights. Several other contemporary commons exist and flourish in the northern regions in Ghana in Africa, as recent research has shown (Kwoyiga, 2019; Nyantakyi-Frimpong, 2020). Likewise, as other recent research has shown, the Swedish Alps provide additional examples of commons today (Head-König, 2019; Schläppi, 2019). So do some religious and Indigenous land, as well as the global commons.

Conclusion

The question of whether the commons is freely available to all is recurrently debated. For Hardin and many others such a "free lunch" would be a formula for trouble. Indeed, Milton Friedman went so far as to claim that there is no such thing as a "free lunch." Privatizing the commons – whether, via Hardin's beaten path or through Ostromian clubs, then, is seen to be the only panacea. We have seen how Henry George took exception to these positions and how the Georgist conception of the commons is consistent with the notion of African commons.

Indeed, looking at African institutions, it is an error to consider freely available to mean a free-for-all, no rights/obligations canvass, or, even worse, no person's land. The Aborigines' Rights Protection Society, for example, strongly argued that there was such a thing as a good free lunch. Indeed, for Africans (see Asante, 1975; Kea, 2010; Obeng-Odoom, 2021), holding land as a commons means that all land has an overall owner: the community. And yet land does not have to be exclusive to only the community members. Strangers are allowed to use land, guaranteed to be compensated for their labour, and required to observe community-devised rules – all without considering land as private property or even a club good. Shared land/labour institutions such as *karafo* are dynamic, ensuring that static conceptions of land are avoided, as they enable strangers to become part of landholding groups, a dynamism unknown to Georgists. The *abunu* and *abusa* reward systems introduce nuances to a flat Georgist tax, as these institutions consider the nature of strangers' contributions to land while acknowledging the position of members of landowning communities. Like George and Georgists, Africanist conceptions of the commons hold that havoc would be unleashed for society, economy, and environment in the event of privatizing land, denying labour of its due reward, and managing the commons from the top down.

Whether these claims can be empirically verified requires additional analysis, focused on addressing the following questions: what happens to society, economy, and environment when land is privatized? How could Africans share their commons, how have Africans lived in their commons in the past? In what ways could the experiences of Africans inform a new ecological political economy? The remaining chapters in this book address these questions.

PART C

The Proof

Chapter Four

Cities

Introduction

Cities face pandemic socioecological crises that threaten their future. Every year, urban pollution from coal-fired power stations are "responsible for the equivalent deaths of more than 2,200 people in South Africa ... and thousands of cases of bronchitis and asthma" ("Eskom Emissions," 2018, p. 6). Emissions in the Global South, more generally, appear to be on the rise. In urban China, for example, the current levels of sulphur and nitrogen dioxide in the air are four times as high as in 1990 (Lu & Liu, 2016). Annually, urban air pollution causes more than four million premature deaths ("Our Urban Future," 2020, p. 111). Solid waste management is another aspect of the urban socioecological crises. Plastic discards blown by wind or washed by running water from cities to oceans constitute 60 to 80 per cent of ocean debris (Mitchell, 2015, p. 79). Urban streams are polluted by about 50 billion plastic bags produced and used annually (Oyake-Ombis, 2012). Urban flooding, which often disproportionately affects areas occupied by the poor, is yet another aspect of the urban socioecological crises, as is the loss of parks, street-side trees, and urban gardens (Manji, 2017; Nagendra, 2019; UN-HABITAT, 2007). These problems can be seen in terms of the lost plant, animal, and human lives, as well as the biodiversity loss.

According to the Conventional Wisdom, to loosely use J. K. Galbraith's (1958/1998) well-known concept, these problems have arisen precisely because of the "tragedy of the commons" (Hardin, 1968). As open spaces, cities, planted on common land and usually free for everyone, in essence are also open to be polluted. Urban residents tend to consider cities as cesspools for their discards and think of the air in cities as something that can be abused. Urban economists of a neoclassical orientation (see, for example, Khan, 2010; Squires, 2013) contend that such urban ecological problems are, in fact, products of externalities, the idea that anything that is external to the market or is not properly priced is overused and abused. Others who are strongly interested in technology point to technological backwardness as the cause of these problems (see additional discussion in chapter 5).

The holy grail of this Conventional Wisdom on the crisis of the urban commons, then, is a spatial trilogy of a tragedy of the commons, externalities, and technological problems, which lead to three sets of distinct, but interrelated, policy choices. Technological modernization (e.g., the use of green plastics) is the expressed preference of technology advocates, while economists tend to prefer internalizing the problem through marketization/getting the prices right (e.g., fees and charges; taxes; removal of subsidies on fuel; or incentives, including payment for good environmental behaviour) to create (1) income effects, (2) substitution effects, and (3) other adaptation effects, such as relocation (for a detailed discussion, see Khan, 2010; Stevens, 2002; Stilwell, 2011b).

The policy alternatives to the "tragedy of the commons" have generated discussions that are both heated and animated. For members of the so-called Wise-Use Movement, new institutional economists, and neoliberals generally, the answer is simply to privatize the city and its resources (Jacobs, 1995, 2020). Land is often the focus for such analysts. The range of assumptions that underpins this alternative is discussed at length in chapter 7.

However, Elinor Ostrom's position on these alternatives requires immediate attention. Leaving it until chapter 7 could create confusion about her place in the debate and impede the unfolding analysis in this chapter. Asking whether Ostrom's work is part of the Conventional Wisdom is moot. She approved of the market-based line of analysis and endorsed privatization, but insisted that it does not need to be corporatized. She tried to show a "third way" of organization in the form of the urban CPRs. In this approach, which she calls "polycentric governance," urban policy should simply endorse a broad sweep of equally valid approaches (see, for example, Ostrom, 2010b, 2012a). This approach has been called "the Ostrom Social Ecological Systems Framework" (Nagendra, 2019, p. 184).

The question, then, is not whether Ostrom's work is part of the Conventional Wisdom. Rather, what are the details of Ostrom's conventional analysis of cities? This question is important because urbanists have not satisfactorily pinpointed what Ostrom meant by the urban commons (for a general discussion of debates on conceptualizing the urban commons, see Huron, 2018, pp. 3–65). The reason – exemplified in the book *Urban Commons* (Borch & Kornberger, 2015) – is that researchers have tried to deduce Ostrom's concept of the urban commons from her generalized

IAD rules, as discussed in chapter 3. Taking a rather circuitous approach, this existing body of literature seeks to *reinterpret* Ostrom's research in terms of the "urban" instead of directly engaging Ostrom's main and direct contributions to urban research. A more direct approach could be to simply engage Ostrom's own work on the urban commons, starting from her PhD work. Then, Ostrom was a scholar grappling with what Vincent Ostrom, her teacher and husband in later years, called the contending approaches to urban governance, ranging from the "consolidated" approach and urban governance, to the two-tier solution, and then to the preferred approach of the Ostroms: the public choice alternative (later called the urban "common-pool" alternative) (Bish & Ostrom, 1976; Ostrom, 2010c).

Approaching Ostrom's work in this way makes it clearer that she regards specific urban organizations such as gated housing communities, informal economies, and slums as urban commons (Guha-Khasnobis et al., 2006; Ostrom, 2010a, 2010b, 2010c). This interpretation of Ostrom's approach to urban commons is the more accurate, but it is often neglected (see, for example, Sarker & Blomquist, 2019). As noted by Mark Pennington (2012):

> Few would suggest that the condominium associations and private (sometimes gated) communities that have spread rapidly over recent years ... as bottom-up alternatives to the municipal provision of collective goods are not a form of "privatization." Indeed, many leftist/social democratic critics have condemned them as such ... These are, however, precisely the type of ... arrangements that Ostrom thinks can and should be used much more widely. (pp. 40–41)

Ostrom does not begin and end her analysis of cities with gating. She also analyses informal economies. As noted in Guha-Khasnobis, Kanbur, and Ostrom (2006, pp. 3–4), "discussions of the formal and the informal have been enriched considerably by the literature of the past two decades on (self) organization of common property regimes." This reading of the commons in cities has eluded urban planners. Thus, contributing to *Planning Theory and Practice*, Libby Porter (2011, p. 117) concludes that "commons property, particularly in urban settings, is almost invisible in planning theory and practice." Porter then goes on to explain what the urban commons is, drawing on Ostrom. In Porter's words, "The urban commons can be seen in community gardens, land trusts, squatting, and common interest developments (such as gated communities). Our contemporary arguments about privatization of city spaces (such as parks and public squares) hinge on often unstated assumptions about commons property" (2011, p. 117).

The connection between Ostrom's work on the commons and research on informal economies is partly that many CPRs were previously regarded as "informal" (Guha-Khasnobis et al., 2006, p. 2), partly because the governance of informal economies constitutes one layer of polycentric governance (Ostrom, 2010a, 2010b, 2010c), and particularly because informal economies are, according to Ostrom, successfully governed by the actors themselves. As Ostrom and her colleagues note, "This is illustrated, for example, by the detailed empirical work showing the highly structured interactions within groups that manage common-pool resources, far removed from any interaction with official governance" (Guha-Khasnobis et al., 2006, p. 6).

These urban commons are expected to make their own contribution to realizing a clean and green planet (Ostrom, 2010b, 2012a). From what Ostrom consistently called a "polycentric perspective" (for a review, see Aligica & Tarko, 2012, 2017), the emphasis on global and national programs to reduce global emissions and mitigate environmental problems has tended to miss how the urban commons and their inhabitants can individually and jointly take simple steps to help address the ecological crisis. Ostrom's preference is for the use of environmental taxes together with a change of behavioural patterns among individuals who reside in the urban commons. It is these strategies that she calls "polycentric systems for coping with collective action and global environmental change" (Ostrom, 2010b). At their core are the urban commons, what, on the day of her death, she called "green from the grassroots" (Ostrom, 2012a). Disciples have continued in this tradition (e.g., Nagendra, 2019). Yet they often do so without recognizing the origins and implications of claiming that their work is inspired by Ostrom's. So, some clarity about the origins of Ostrom's analysis of the urban commons is needed.

She took her concept of the urban commons from Charles Tiebout, who coined the related idea of "voting by the feet" (Tiebout, 1956). This idea emphasizes how urban residents, dissatisfied with their local governments, respond by exiting the city or parts of it. The idea is analogous to Milton Friedman's preference for self-governing spaces where individual urban residents reject metropolitan Keynesianism and live in areas where they make their own rules (Peck, 2015, pp. 130, 135). As Albert Hirschman (1970) showed, this strategy is distinct from "voice." Typically, this value signals the prospect of cooperation with others, including state officials and other citizens. Instead, Ostrom's idea of the urban commons is steeped in the notion of consumer sovereignty. A core value of the Conventional Wisdom, this orientation is suspicious of governments.

In his contribution to *Public Choice*, Klarita Gërxhani (2004) throws more light on the public choice approach to analysing the informal economy.

Starting from the assumption that all agents are rational utility maximizers, public choice analysts question the assumption that governments are honestly seeking to support the public good (Gërxhani, 2004, pp. 286–287). Governments seek their own re-election, which entails taking certain actions that may not necessarily be in the interest of the public. In turn, public choice analysts endorse informal economic arrangements as rational alternatives. Not only do they avoid corrupted state processes, but these informal agents also constrain corruption while contributing positively to economic growth (Gërxhani, 2004, pp. 288–289). More fundamentally, they ensure that the cardinal principle of consumer sovereignty remains sacrosanct (see Desmarais-Tremblay, 2019).

Not only did Ostrom accept this line of analysis, she extended it . She sought to show that these individuals *successfully* manage their *commons*, what *The Economist* called "the unplanned economy" ("Destroying the City," 2019, p. 32). Thus, Ostrom must take the coexistence of "slum governance" (Stacey, 2019; Stacey & Lund, 2016), "suburban governance" (Hamel & Keil, 2015) in the form of how the elite and wealthy urban residents govern themselves in gated estates, and the governance of "nature in the city" (Nagendra, 2019) as prima facie evidence of success. By seeking to show that small communities succeed in managing themselves, she also sought to extract the general rules about how small communities successfully self-govern without the intervention of the state. This emphasis, then, was an extension of the public choice approach to urban studies of which Vincent Ostrom was a pioneer. Emphasizing how global (urban) socioecological problems could be addressed by the sum-total of various rational individual actions taken at a variety of scales became Elinor Ostrom's public choice approach to the urban commons. According to her, "[s]ustainability at local and national levels must add up to global sustainability. This idea must form the bedrock of national economies and constitute the fabric of our societies" (Ostrom, 2012a, n.p.). Ostrom extended this public choice approach from its traditional focus on politics and government to the wider spectrum of governance (Boettke, 2010), an extension that simultaneously widens the skepticism about the state and increases the faith in individual rational decision making. Is the Western Left Consensus on the urban commons different? While the Conventional Wisdom neglects questions of justice, the Western Left Consensus is centred on justice. This emphasis is important because

> To say that the city is a commons is to suggest that the city is a shared resource – open to, shared with, and belonging to many types of people. In this sense, the city shares some of the classic problem of a *common-pool*

resource [emphasis added] – the difficulty of excluding people and the need to design effective rules, norms and institutions for resource stewardship and governance. (Foster & Iaione, 2019, p. 237)

Defending the idea that the city belongs to all is, therefore, fundamental to the Western Left Consensus. If the tragedy of the commons, in fact, arises from the capitalist logic of privatization itself, the argument goes, the solution must be commoning everything, especially the product of labour, including urban technology. A major urban commons group is the Co-Cities Project, (http://www.commoning.city), led by Sheila Foster. According to her and her colleague, Christian Iaione (Foster & Iaione, 2019, p. 236), "The goal of this research project is to enhance our collective knowledge about the various ways to govern urban commons, and the city itself as a commons, in different geographic, social and economic contexts" (Foster & Iaione, 2019, p. 236). The effort here is to common "community gardens, parks, neighborhoods … and urban infrastructure such as urban roads … and later jointly to conceive of the whole city as a commons" (Foster & Iaione, 2019, p. 236). This is the idea that Dan Webb (2017) has recently characterized as "open cities." These are spaces to which everyone in the city has a common right. This idea might appear antithetical to the Conventional Wisdom, but the two positions can be strikingly similar.

The Western Left Consensus shares grounds with the Conventional Wisdom in three areas. The first is the reliance on Ostrom's concepts, such as CPRs (see my italics in the Foster & Iaione, 2019, quotation on p. 89). The second is the wholesale application of Ostrom's IAD to analysing the urban commons in the Global South while still retaining some residual interest in justice (see, for example, Adamu, 2012; Mundoli et al., 2019). Third is the static and romantic view of the ultimate agency of urban informal residents held by both the Conventional Wisdom and the Western Left Consensus. This overlap can be seen in the work of leading progressive scholars such as David Drakakis-Smith (1987) and Asef Bayat (1997). The wide range of examples used by Stavros Stavrides (2016) in *The City as Commons* (see especially pp. 129–158 of Stavrides's 2016 book) could be taken as evidence of the overlaps between the Conventional Wisdom and the Western Left Consensus. More specifically, on the agency of the informal urban commons, Kim Dovey (2012, p. 349), for example, notes that "this may be, for some, simply an image of poverty or underdevelopment, but it is much more one of entrepreneurial flexibility, adaptation and creativity." Indeed, for Dovey (2012, p. 364), "Informal settlements have arguably the lowest carbon footprints of any form of urbanism on the planet – they often utilize recycled materials at high densities with low-rise morphologies, close to employment with very low car dependence."

For Donald Brown and his colleagues from the International Institute for Environment and Development (Brown et al., 2014), urban policy can usefully focus on encouraging "urban informality and building a more inclusive, resilient and green economy." This argument – effectively claiming that informality is sustainability – is long-standing. John Turner famously argued in the 1970s that "with a vastly greater range of lightweight, low-powered, potentially decentralizing technologies the possibilities of effective action by local groups and associations, and of rapid general change, are vast and immediate" (Turner, 1976, p. 9). Development planning has a long history of research on looming urban disasters and hazards, and the resulting "adaptation" of cities through the dynamism of what Ostrom would later call urban commons or urban CPRs (for authoritative reviews, see Bryceson, 2016; Clay, 2017; Morgan, 2018).

The account of Matthew Khan, an urban economist trained by Milton Freidman, shows the orthodox roots and essence of the jarring literature on disparate ideas such as urban *vulnerability, techno-fixes, privatization, pricing urban resources, adaptation,* and *livelihood strategies.* As systematized by Matthew Khan (2010), putting a price on environmentally destructive products would enable self-interested consumers to make rational choices between environmental bads and environmental goods. Self-interested individuals would reduce how much they spend on expensive environmental "bads" (income effect), which these calculating individuals would replace with cheaper environmental "goods" (substitution effect). The sum of several self-interested individual actions, including those in the urban common pools, that is, widespread individualism, would force markets to shift in favour of green production, as capitalists try to produce and sell green goods to make more profit. It is through these processes that capitalist cities self-correct their socioecological crises.

From this background on the debates about the urban commons, three questions arise. First, do the urban CPRs arise as a rational *individualized* alternative to address the crisis of the state? Second, could the activities of the urban commoners address the urban socioecological crises? Third, what accounts for the crisis of the urban commons, whether defined in terms of the Conventional Wisdom (parks, gardens, roads, bridges, squares) or the Western Left Consensus (the right of everyone to access and control the city)?

Based on evidence personally collected from cities in Africa and experiences analysed by others, I argue in this chapter that for all these questions, urban reality is far more complex. Markets have not deterred polluting behaviour; the rich and mighty continue to pollute. Marketization is ethically problematic because it implies that the rich can destroy what is our commons, so long as they have money. Private property in nature can also be said to be the cause of the "tragedy of the commons." What drove the overuse and pollution problem in Garrett Hardin's account can be said to be an individualistic logic – which helps to explain the pollution of "private" mining sites or the pollution of the commons (e.g., rivers) by private drilling/mining activities on adjoining land (this argument is further developed in chapter 7). Technological modernization could create even more environmental problems, what is usually called the "Jevons paradox" (this argument is further developed in chapter 5).

Polycentricity through urban common pools is similarly problematic. Not only does it ignore uneven and combined urban development, but urban CPRs, especially informal economies, also grew from oppression, not freedom. They are maintained by suppression, not liberation, and producing these urban commons has institutionalized segregation, destroyed natural environments, and curtailed the rights to urban land. The idea that, through market instruments, individual responsibility, and privatization, the urban socioecological crises could be addressed overlooks the structural causes centred on the monopolization of land.

To demonstrate this argument, the rest of the chapter is divided into six sections. Starting with "Evictions and Enclosure," the chapter shows the contemporary fraught relationship between the state and informal economies. "Neoliberalism and the Privatization of Urban Space" questions Ostrom's idea of the urban commons by demonstrating that those residing in the urban commons may be dissatisfied, but their discontent is with the market and the marketization of the state in ways that contradict Ostrom's public choice approach. Then, the very concept of "markets" is problematized under "Neocolonialism and Western Planning Models," which shows that, in contrast to what political economists usually claim, markets and neoliberalism are actually patterned after a previous, bigger force. This force, as the next section on "The Persistence of Apartheid: Planning as a Wedge" shows, does persist and creates a path-dependent social problem. "Structural Limitations of 'Green from the Grassroots'" shows the limits to the agency of environmental informal workers by highlighting various aspects of work and waste and how colonialism corrupted the idea of "land," promoted wage labour, and monopolized capital. "The Tragedy of Privatizing Nature" looks at the socioecological crisis that arises from privatizing nature.

This social change institutionalized the many urban socioecological problems in cities today, although this finding eluded Ostrom and has continually baffled her followers and the advocates of the Western Left Consensus.

Evictions and Enclosure

Informal market centres in Africa have long been subjected to disruption in the name of planning and orderly development. "On Saturday, August 18, 1979," wrote Claire Robertson (1983, p. 469), the American economic historian, "Makola, the Queen of Accra markets, died. It was an ignominious death unfitted to one so alive. The bulldozers arrived after the soldiers had plundered money and goods from the stalls and flattened the area to rubble." According to the state newspapers, this process led to a "happy tragedy" that produced "tears of joy" for the "worker, the common man." Fast forward to 2007, a quarter of a century later:

> On October 19, 2007, at about midnight, a team consisting of some members of the Ghana Police Striking Force and specially trained security guards from the Accra Metropolitan Assembly (AMA) arrived at the Tema Station, a bus terminal in the Greater Accra Region, and carried out a "decongestion exercise" which is, razing down all "unauthorized" structures, including stalls and kiosks erected without first obtaining planning permits. (Bob-Milliar & Obeng-Odoom, 2011, p. 264)

Similar evictions have taken place in Harare, Zimbabwe, Tunis, Tunisia, Lagos, Nigeria, Bameda, Cameroon, Dar es Salaam, Tanzania, Kigali, Rwanda (Campbell, 2014; Manirakiza, 2014; Ojong, 2011), and many more cities in Africa. In 2009, about eight out of every ten African countries tried to reduce the population in its cities using evictions, watered-down eviction programs, and many other social interventions (UN Department of Economic and Social Affairs, 2010). In all these cases, the evictions involve people who are socially, spatially, and economically marginalized and excluded. They tend to be in an imaginary legal space called the "informal system." It is legally imaginary because it can sometimes dovetail seamlessly into the formal system because it serves the formal system even as the formal system forces it to remain underdeveloped. Such evictions are costly to the state, devastating to the individuals affected, and destructive to large communities. Some individual eviction events cost the city authorities in the Ghanaian cities of Accra and Kumasi about $100,000. For financially stressed cities in Africa, this is a large sum.

For the public, physical assets have been lost, social networks broken, and savings obtained from long hours of precarious work lost through the destruction of assets (Obeng-Odoom, 2011c, p. 371). Such evictions are also accompanied by emotional, social, cultural, and economic repercussions. In cities such as Lagos, where social networks have been deemed particularly strong (Asante, 2020; Asante & Helbrecht, 2018; Lawanson & Oduwaye, 2014), there are compelling grounds to expect massive psychosocial pressures from evictions.

In Bameda, Cameroon, evictions deprived some 700 people of premises for small-scale trading. It also deprived the city's urban farmers of food to eat and to exchange, as the city authorities decided that farming space was needed for the planting of flowers. In the case of business premises, the evictions were justified as a way to make room for upmarket shops that give the city a modern and clean appearance. Such deprivations lead to inequality as the "new" spaces are taken over by the rich, and the evicted poor struggle to find non-existing alternative livelihoods (Ojong, 2011, 2020). One study found that of all those evicted in Kinshasa, Democratic Republic of the Congo, 80 per cent had to find homes in other spontaneous settlements either in the same or worse conditions (Manirakiza, 2014). In Dar es Salaam, Tanzania, one evictee sums up the frustration as follows:

> I don't feel like the government has valued me. I have been living here since 1973, that's a very long time. The government watched us when we built our homes. Now they are just telling us to leave, but they are not giving us any place to go. Nowadays, building materials costs are much higher than when I built the house. People who have already moved from here, their compensation was very low compared to the value of their home. I have been offered money but I am staying because the price was undervaluing me. (Campbell, 2014, p. 198)

Viewed ahistorically, these are examples of the brutal African state, unenlightened and uneducated, corrupt, undemocratic, and unsympathetic. When the then Zimbabwean president marched the powers of the state against marginalized people in the infamous Operation Murambastvina ("get rid of trash"), Jack Straw, then British foreign minister, was quick to tell the G8 economically rich and elite countries that "if the reports are simply half true – and we believe them to be much more than half true – this is a situation of serious international concern," and ought to be condemned ("Zimbabwe Clearances Condemned," 2005). According to one pundit, Mugabe's government was "ruthless" and "the situation in Zimbabwe ranks among the world's worst government-created humanitarian disasters." Consequently, this critic called for greater isolation and economic sanctions against the Zimbabwean leader and his cronies (Schaefer, 2007). Given the death and destruction of property, the wails and travails that accompanied this operation, such sentiments are understandable. However, they seem to suggest a dichotomous framework of bad Africans and good Westerners, which encourages belief in Western responsibility to "civilize" or enlighten the Africans. American political scientist Michael Bratton has consistently made this argument, suggesting that failures in Africa are embedded in the African institutions of today and that authoritarian regimes are in the nature of Africans, especially those south of the Sahara. Westernization and capitalism, according to this analysis, might eventually export democracy to Africa (see, for example, Bratton & van de Walle, 1994; "How to Beat," 2020; Mallaby, 2010).

Neoliberalism and the Privatization of Urban Space

If we dig deeper, we see that these contemporary evictions are but a reflection of a logic whose roots are deeper and getting deeper in the current dominant economic paradigm of neoliberalism. Although quite vague in principle (Dunn, 2017), when viewed in context and along with specific theoretical propositions, neoliberalism can be analysed consistently. In Africa, its nature differed in two broad epochs. The current epoch (ca. 2000 to date), characterized as the era of "good urban governance" often playing out as democratization, entrepreneurialism, and decentralization has entrenched a segregationist orientation, according to which caring for cities divides them along commodity and class axes, disguised as "governance for pro-poor urban development" (Obeng-Odoom, 2013b, 2020). Many urbanites have tried to protest this trend (Asante, 2020), but the role of urban planning has been subcontracted to expatriate planning consultants and estate developers who are not based in Africa and who imagine the continent from afar – a phenomenon South African planning scholar Vanessa Watson (2014) has called "urban fantasies." While some of these fantasies have fulfilled some local aspirations (Bhan, 2014), their overall effect on cities in Africa is to segregate or reproduce and intensify segregation inherited from the colonial period (Cain, 2014). They are not intended to solve the "real" tensions of city life faced by the majority poor in Africa. Instead, they consider cities in Africa as the defining unmanaged commons or development frontier (Côté-Roy & Moser, 2019), where private property *must* be imposed to address the tragedy of the commons.

In the context of this Conventional Wisdom, even where these externally produced plans solve real problems, their emphasis is on papering over deeper cracks. Examples of such outcomes can be found in the many Chinese projects in cities in Africa. Much of this investment draws on imported Chinese labour that, incidentally, tends to spend very little in the cities in Africa. While China is seen to have been magnanimous in providing infrastructure in cities in Africa, it has been argued, in the case of Ghana, that such investments play the role of physically asserting China's power and influence in Africa. Chinese architecture, such as the National Theatre of Ghana, according to the architect who designed it, seeks to project Chinese culture (Amoah, 2014, p. 8). Of course, it simultaneously seeks to facilitate the ongoing neoliberal modernization in the country, too.

There have been limited linkages in those cities hosting such investments. Even worse, the sale of Chinese products, or counterfeit versions of African products developed by Chinese producers and sold more cheaply in urban markets in Ghana have contributed to the death or decline of local industries, such as textiles, and, hence, have opened another avenue of inequality (Axelsson, 2012; Obeng-Odoom, 2020). The Senegalese and Malagasy urban residents have had similar experiences with Chinese investment (Cissé, 2013; Obeng-Odoom, 2020; Tremann, 2013). Coupled with growing transnational activities, including transnational housing forms, a new Accra, or what Richard Grant (2009) has called "a globalizing city," has emerged with its growing gated housing communities and related ballooning rents and rent-generated segregation. Even if Chinese interventions and investments retain distinctive features, compared with Western neocolonialism, these experiences suggest that unbridled faith in the "Chinese model" is naive.

In an earlier epoch (1980s–1990s), planners adopted an explicitly anti-urban sentiment in development policy (Obeng-Odoom, 2013b, 2021). During this time, cities were seen to be benefiting at the expense of the country in what came to be widely regarded as "urban bias" (Lipton, 1977). This "urban bias" was seen to be part of a bigger problem of "over regulation." In turn, Western powers decreed and pushed for the removal of labour from state payrolls and a general reduction in public sector employment, which was predominantly located in cities. This displaced labour, discarded for no fault of theirs, landed on the African streets without work (South African Population Research Infrastructure Network, 2002). Paradoxically, the urban population was growing around the time but with persistent lack of investment in cities, the development of large slums in Africa was quite inevitable – a process quantitatively demonstrated by Dr. Sean Fox (2014) of University of Bristol (see also Fox et al., 2018) . There was the policy of promoting "global cities," which entailed preferential treatment for modern, expatriate, business-friendly cities that tended to promote the "clean-up" or "decongestion" of the African streets to welcome business or to ensure the free movement of goods and services (Obeng-Odoom, 2011a, 2011b, 2011c, 2016b, 2020). In this way, it was possible to treat "cities as engines of growth," to "enclose the commons," and further evict or convict those "trespassing" on the resulting private spaces. Neoliberalism, then, is one major intensifier of the urban inequality and evictions we see in Africa today.

Neocolonialism and Western Planning Models

If we look at the situation from an even longer historical perspective, the picture becomes even more startling. The evictions of today reflect the politics of the African postcolonial era, particularly the 1960s and 1970s. Post-independence governments aimed at demonstrating to the colonizer that they could do the same things that used to be done in the colonial era. The intention was not to subjugate but rather to show that Africans had the same capacity as the colonizers to run the postcolonial nation in the same manner as the colony had been run. Massive projects of modernization were undertaken. In Ghana, shopping malls and large monuments were constructed to display modern architecture. Accra, the capital city, in particular became a showpiece of modern architecture and built form. J. B. Hess (2000) famously noted that "the Nkrumah administration's response to colonial regulation was a distinctive 'imagining' of architectural modernism, an imagining which allied the heroicized image of Nkrumah with a culturally homogenous notion of the 'nation'" (p. 53). Buoyed by consultant architects from Britain and America, Africans' desire was to show that they could "catch up." As with Mao Zedong's "Great Leap Forward, which was launched in 1957 … to 'catch up with Britain in three years and surpass America in ten years' … Nkrumah's … developmental goals are particularly instructive: 'what other countries have taken three hundred years or more to achieve, a once dependent territory must try to accomplish in a generation if it is to survive'" (Amoah, 2014, pp. 2–3).

A growing spatial divide became evident, however, particularly because of the inherited planning system. A paper published in *Cities* makes this point: "Planning in Sub-Saharan Africa owes much to the colonial legacies that inform the shape and composition of African urban spaces and places. This applies to legislation, institutional systems and planning education" (Odendaal, 2012, p. 174). At the base of the planners' skills is their education and, to date, the 69 planning schools in Africa typically teach archetypes of planning desired by the colonizer and praised as the ideal – even though this ideal is not reflective of what exists. Informality, slums, the disconnect between plans and the capacity to implement them, climate change, and the political roots of planning have all been overlooked or given scant attention in the curriculum and indeed in the deliberations of the professional planning bodies, even though these problems are crying out for urgent attention in the contemporary city in Africa. Retaining supposedly pristine European values has been deemed preferable. The degree of colonial education varies, but in large measure they retain the old order of segregation (Muchadenyika, 2020; Odendaal, 2012). The outcome is not only a colonial but also a colonizing philosophy of education that looks down on everything "African." This paradigm hence, fosters a lingering sense of inadequacy and self-doubt among the African peoples and what they cherish (Nyamnjoh, 2012, p. 129, 2019). Consequently, acting, thinking, and seeing like a colonizer become ideals. From this perspective, the planners froze rules and regulations, some of which were inapplicable to Indigenous ethics and ethos but were consistent with certain European ideals (Konadu-Agyemang, 2000; Njoh & Chie, 2019).

Imitation and mimicry became celebrated in more than one facet of society in the newly independent countries. According to social anthropologist Sasha Newell (2012), "The act of imitation is a matter of national pride in Côte d'Ivoire … Ivoirians were the very best imitators of Europeans … There is no shame then, in being derivative. It is precisely in the ability to imitate with precision that many urban Ivoirians locate their sense of prestige" (p. 1).

Some of the reasons have been identified, but there are others. One is that expatriate consultants were maintained, and expatriate courses introduced, to teach the new civil servants the white people's ways. Also, professional planning bodies sprang up to reinforce the white people's ways. This was hardwired into the Western institutions of planning. Kwadwo Konadu-Agyemang (2000) analysed the fascination with colonial and apartheid urban planning. More recently, Nancy Odendaal (2012) has shown that the professional planning bodies themselves are the cause of the problem, as they have become agents for the upholding of planning standards of which a credo is apartheid planning. Resource constraints, not only of planning schools, as Odendaal shows, but also of planning departments deprive them of the energy to innovate (Gaise et al., 2019; Yeboah & Obeng-Odoom, 2010).

The Persistence of Apartheid: Planning as a Wedge

Going back still further in history, we arrive at the grand structure: the colonial-urban structure in Africa that insisted on a certain logic of order (ca. 1870–1950s). This logic permeated the different phases of colonial settlement: mercantile, industrial, and late colonialism, to use the nomenclature of David Drakakis-Smith (1987). The key difference among these relates to the degree of exploitation, expropriation, and appropriation. Regardless, colonialism fostered segregation: one place for the white colonial edifice; the other place for the black "others." Separation and segregation were hardwired into urban planning. Activities considered to be for the "blacks," the "black system," were shoved out of sight and, in terms of planning, out of mind. While the French used to reserve an area not to be inhabited called *cordons sanitaires*, the European English-speaking colonizer used a non-residential belt to separate residential areas of the "superior" (read "white") from the "inferior" (read "black") races (Mabogunje, 1990, p. 138). There in Africa, strangers were dictating to natives where to live on their own soil.

The colonists justified such segregationist policies on grounds of science, that is, that integration with Africans would lead to malarial and yellow fever infections among the colonizers. Yet scientists from the colonial metropolises themselves such as Carlos Finlay, Philip Curtin, Alfonse Lavernan, and Ronald Ross demonstrated conclusively that the Africans were not the vectors of malaria: the anopheles mosquitoes were (Njoh, 2009, pp. 10–11). Indeed, if anything, there was evidence that sexually transmitted infections, such as syphilis, previously unknown in Africa had arrived on African soil (Tsey & Short, 1995). Yet pundits like Joseph Chamberlain, one-time British colonial secretary in West Africa, insisted on separation to guarantee the health of the whites (Njoh, 2009, 2014; Njoh & Chie, 2019). Raw racism was injected into the planning system not only in the form of separation into "white" and "black" quarters, but also in terms of offering lavish health support and facilities, such as hospitals, subsidies for healthy foods, medical supplies, and booklets for healthy lifestyles for the whites (Njoh, 2009, 2014; Njoh & Chie, 2019; Ojong, 2020; Tsey & Short, 1995).

In the urban centres of Kumasi and Sekondi-Takoradi in the then Gold Coast, massive health infrastructure was constructed, but this served the needs of only the expatriate staff and the few African elites. The majority of Africans had no health support. Spatially, the 1919 guidelines on *Residence in the Segregation Areas of the Gold Coast* made it illegal for African children to go to the white areas. Also illegal was the sale of title of "European lands" to Africans. Two years after the publication of the guidelines, the Takoradi township plan was also published. A key feature of the plan was that it created three distinct settlements: one for the colonizers, another for the elite Africans, and the third for the general blacks. Invariably, Africans were evicted and the best parts in the city were reserved for whites (Obeng-Odoom,

2014b, 2020; Tsey & Short, 1995). Common land was turned into symbols of compartmentalization. Common parks and pasture in Accra and elsewhere in urban Africa – indeed, much urban green space in Africa – was used as buffer to separate the colonists from the "other" (Arku et al., 2016; for additional specific details of diverse societies in Africa, see Obeng-Odoom, 2013a, 2013b, 2013c, 2013d, 2013e, 2020).

Urban governance in the postcolonial and neoliberal epochs explains much of the invisibility of common parks in many cities in Africa today. Such commons have been sold to property developers for the construction of gated housing estates. However, the colonial use of parks for suppression of peoples of African descent is part of the story, too. Together, these forces help to explain the disdain for, and absence of parks in, much of urban Africa today. A few parks exist in cities such as Accra, but for the most part such commons are inaccessible because the authorities charge user fees, which only the rich can afford. Some low-income users can pay the user fees, but the opportunity cost of doing so is prohibitive. The urban commons have become de facto enclosures (Arku et al., 2016), while the remaining commons are used as burial grounds for the rich and mighty (Arku et al., 2016). For the most part, only marginal lands remain as the available "commons" for the many urban poor.

These inequalities were systematically carried out even under the French so-called "policy of assimilation" (Njoh, 2009, 2014; Njoh & Chie, 2019). In Equatorial Africa, in particular urban areas in Brazzaville, enclaves were created for Europeans to protect them from Congolese who were forced into inferior spaces, such as Bacongo and Poto-Poto, under contested claims that European culture and African culture were immiscible, that the Africans were vectors of disease and were noisy, so they were not conducive to integration. In essence, similar policies were adopted in the Guinean capital, Conakry, although in this city, racism was implemented through the market in the sense that housing in European areas was affordable mostly by Europeans; if Africans could afford to live there, they had to use European building materials and act "European" to be accepted (Njoh, 2009, pp. 11–12; Njoh & Chie, 2019). Thus, the market was only a camouflage for racism. Indeed, the French – unlike as they had done in France, recognizing common property of the French peasants – refused to acknowledge the commons in colonial Francophone Africa. They insisted on subdividing the commons into private parcels titled to individuals (Tabachnick, 2016), a process that, according to more recent research (Korbéogo, 2018) is ongoing in Francophone and other parts of Africa.

Ostrom's idea that living in informality is a kind of rational decision for self-governance is questionable. Informality is a complex socioecological feature of the urban and regional development process in Africa. Its form may differ in place and over time. However, the effects of the cumulative privatization of land, coupled with various historical and contemporary forces, have shaped the formation, form, and transformation of informality. Without considering the ramifications of enclosing the land commons, therefore, the Conventional Wisdom, especially the work of Ostrom, is highly problematic, indeed misleading, and a gross oversimplification of informality.

The question, then, is no longer how informality arises, but to what extent it constitutes a compelling solution to the urban ecological crisis. According to the Conventional Wisdom, not only is the rise of informality benign in terms of the autonomy it gives to its participants, but informality is also a solution to socioecological crisis. Described by Elinor Ostrom (2012a) as "green from the grassroots," this informality-as-panacea idea claims that informal waste picking, for example, can address the planetary problems of plastic waste in the Global South (see also Nagendra, 2019). This claim requires empirical analysis to ascertain its validity. Specifically, what is the nature of the work involved in "green from the grassroots"? Under what conditions does labour work in this approach? Can this model of green grassroots address the planetary problem of (plastic) waste? If so, how and, if not, is the model also neocolonial?

Structural Limitations of "Green from the Grassroots"

Abidjan is a useful case study for addressing these questions because its experience illustrates the place of cities in generating and maintaining the current age of plastic waste. Once compared to Paris and Geneva (UNEP, 2015, p. 6), a rather neocolonial way of assessing cities in Africa, this city in Côte d'Ivoire, West Africa, is, nevertheless, now engulfed in plastic discards, contributing to what Anderson (2015a, p. 139) called "one of the world's ugliest problems." Being an eyesore is obvious enough, but the socio-economic effects of plastic waste require more careful analysis. Tourism and fishing as economic activities risk decline because of the extensive pollution of water bodies with waste (World Bank, 2015). Plastic waste in the city tends to make its way into inland water sources. Consequently, this waste limits the amount of freshwater available in the city; blocks drains, making the city vulnerable to flooding; and generates the conditions for mosquitoes to breed. Plastic waste also interferes with life in water bodies, for example, through stifling the growth of nutrients required for water organisms. It has been estimated that, annually, 4.4 million m^3 of waste pollutes the Ebrié Lagoon (Komenan, 2010), Abidjan's major lagoon.

The World Health Organization (WHO) recently found that the Ebrié Lagoon has been so heavily polluted that it has become an inhospitable habitat for fish (World Bank, 2015, p. 88). Plastic constitutes an increasing share of waste in Abidjan, rising from 6.99 per cent in the 1990s and early 2000s (Ministere de L'Envionnement et Du Cadre de Vie, 2001) to current levels of 11 per cent in high income private homes, 7 per cent in low-rent settlements, 7 per cent in spontaneous settlements, and 8 per cent in poorer shared housing (Andrianisa et al., 2016). Ebrié Lagoon remains navigable, but at the shore, plastic waste chokes off so much water that water flow was difficult, creating the conditions for mosquitoes to breed in the lagoon (World Bank, 2015). Globally, this problem is also linked to the "micro plastic" issue (Anderson, 2015a, 2015b) or the environmental problems occasioned when plastic waste that disintegrates into smaller, invisible particles poisons life in water bodies and destrosy terrestrial life as well.

Much urban informal labour has developed to clean up the crisis. In December 2015, I conducted fieldwork to find answers to questions about the nature, conditions, and consequences of green from the grassroots in Abidjan. I had previously conducted similar research in West Africa (Obeng-Odoom, 2014d), so I was familiar with the debates on this labour question, how it plays out in cities in Africa, and their continuing struggle with neocolonial processes.

The Nature of Green Work

The groups of informal workers – often called "waste pickers" – can usually be seen on dumpsites, around hotels, and in other places where plastic waste can be found. In Abidjan, sometimes pickers buy plastic waste from prior collectors such as room service workers who collect waste from the hotel rooms they have been contracted to clean. These pickers, in turn, sell to (1) market agents in Adjamé, (2) directly to traders such as those who sell fruit juice, and (3) to recycling companies. How much pickers are paid depends on how the waste they have collected is valued. Generally, pickers are paid either based on the weight or on the quantity of the plastic waste they collect. Three big waste bottles fetch US$0.16, while four small bottles of waste fetch US$0.16. Plastic waste bags (sachets) are increasingly becoming valueless, as they are not usually in demand.

As a complement to another job, waste picking may be rewarding to the pickers and their families because it brings them extra income. However, as a main source of livelihood, waste picking generates meagre income even by Ivoirian informal economy standards. Traders, for example, those who sell fruits and fruit juice in the informal markets of Port de Fruits in Plateau will make more money as they buy their "raw materials" (plastic bottles) cheaply from pickers to package their home-made fruit juice to sell.

Apart from pickers in the city, there are also pickers on the dump-site on the outskirts of the city. In Abidjan, the largest waste dump is Akouédo, a privately managed for-profit landfill active since 1965, oper-ated by the private entrepreneur Pisa Impex, with some supervision from the Mission for the Conduct of Municipal Operations. Between 200 and 1000 people scavenge the dump, most of whom are women (UNEP, 2015). For many of these women, the dumpsite doubles as their home. During fieldwork, I observed mothers working on the dumpsite with their babies strapped on their backs.

The nature of this work requires further description. According to Brechbühl (2011), most of the pickers here are from the "outcast" groups in Côte d'Ivoire. Migrants from the North of the country and others from neighbouring countries, notably Mali and Burkina Faso, dominate this class of workers (for similar experiences in other cities in West Africa, see Obeng-Odoom, 2014d). Men on the dumpsite literally dive into approaching dump trucks to collect the best/most valuable garbage. Working with no protective gear, masks, or boots, the pickers often are injured by sharp objects such as broken glass and they inhale toxic fumes. The private waste companies that deposit waste in the landfill do not sort the waste, so women tend to do three types of work here (Brechbühl, 2011): collecting (gathering from the dump); upgrad-ing (sorting into waste types); and selling (to other pickers or to agents for sale to industries or markets for reuse).

Slightly better off women do upgrading but, generally, women intro-duce their children to all aspects of the work. With the money they earn, Brechbühl (2011) shows that women tend to either restrict (deprive themselves of the most basic necessities of life to save), support (spend the money they make on vulnerable families), or invest (mostly in the primary education of their children). The work of these pickers is also highly gendered and generational. Brechbühl's (2011) analysis provides even more graphic details, although it tends to emphasize exploitation of women.

Conditions of Green Work

In the field, I observed both some collaboration and work division. Per-haps these social relations recall Francis Nyamnjoh's (2015) attempt to frame social change in Abidjan in terms of *ubuntu-ism,* the idea that being human in Africa means supporting other humans, a kind of "I exist because you exist" idea. So, in Abidjan, some women may till the land to supplement their incomes – an exercise in which men are visibly involved. The tension is that the soils in and around Akouédo are so polluted (UNEP, 2015) that yield must be low and the product can only be toxic, highlighting the difficult tensions between *ubuntu* (aspirations to support one another) and *ubuntu-ism* (structural tensions and contradictions in the wider economic system both locally and globally) that Nyamnjoh (2015) describes in his own study. In my own field observation, the food sold on the dumpsite is also prepared in an environment with suffocating stench emanating from a mixture of many toxic substances. Explosions and fires are common. The most well-known is the infamous Trafigura Ltd case, in which Trafigura, a transnational corporation, conspired with a local Ivoirian company to dump toxic waste in the dumpsite. In what was called a "corporate crime," Amnesty International found that, as a result of this waste, 100,000 people in and around the landfill site sought medical attention, while 15 people lost their lives. The pictorial evidence presented by Amnesty International (2015, see front and back cover of the report as well as pages 1 and 5) showed that women suffered more from this environmental crisis. Either way, waste picking is usually a poorly remunerated, highly exploitative, and harmful informal activity which benefits mostly industry, to which most returns go.

Similar findings have been reported elsewhere in Africa. In Pretoria (South Africa), for example, Rinie Schenck and Phillip Frederick Blaauw (2011) and Phillip Frederick Blaauw et al. (2019) have shown that most waste pickers are male, but they are also migrants and black. In South African society, these identities signal marginalization and interiorizations. Although these waste pickers sell to profit-making buy-back corporate enterprises, they earn so little from waste picking (US$21.32 per week) and their earnings are so variable that they can barely support themselves, let alone other family members. Much like the case of Abidjan, the waste pickers live in poor conditions. They have substandard housing with inadequate water and sanitary facilities. Coupled with harsh working conditions, such as long hours of work, pushing heavy trolleys, and the constant exposure to the risk of injuries, these waste pickers have little familial or wider social support.

Support for these waste pickers is clearly and urgently needed, given that they make such an important contribution to making cities cleaner. The question is not whether they must be supported but whether this solution is feasible with all its social costs. This approach has important structural limitations for three reasons (Obeng-Odoom, 2014d). First, the rate by which plastic waste is generated in cities far exceeds the rate at which pickers can collect the discards. Second, the potential health hazard to the pickers would suggest that any attempt to increase the speed at which waste is picked could only accelerate the rate of urban ecocide. Third, as unsold waste tends to be dumped in the neighbour- hood of the poor pickers, waste picking constitutes a mechanism for transferring the discards of the rich who live in gated communities or visit hotels and holiday resorts to poor neighbourhoods.

The case of Abidjan and several other cities in Africa (see also the experiences of Sekondi-Takoradi in Ghana in Obeng-Odoom, 2014d) highlights that Ostrom's posited solution does not and cannot address the crisis of the urban commons. Instead, it has serious implications for the continuing neocolonial urban development process in Africa

Neocolonialism and Green Work

Green from the grassroots has important neocolonial features. One expression of this dynamic is the internal colonialism in the form of the rich and powerful shifting its discards unto the poor and the weak who manage waste under appalling working conditions. Another manifes- tation is at the global level, where green from the grassroots entails a certain new international division of labour in which the Global South cleans the discards of the Global North. Indeed, in seeking to keep their space clean, countries such as Australia, the United States, and Canada have continually shipped their waste to China and the Philippines. Many other countries in the Global North dump their waste in Africa (see, for example, Grant & Oteng-Ababio, 2016), only to turnround and, paradoxically, describe the Global South as polluted.

These local and global manifestations of neocolonialism are inter- linked. The conditions of the global exchange, much like the conditions of local waste exchange, are exploitative. With mixed waste from the Global North disguised as "recyclables," the Global North tries to clean its waste by shifting it to the Global South under poor terms of eco- nomic engagement (Semuels, 2019). Coupled with their disrespect for local institutions, the Global North champions a new international divi- sion of labour that bears clear neocolonial stripes.

Not only do the states involved in subordinate position get disre- spected, but also being manipulated means that the victim countries, in turn, victimize minority populations in their societies. In the end, the affluence and leisure of the Western world ends up as discards to frustrate the emancipation of the South.

The recent spate of "(plastic) waste wars" with many exploited coun- tries in the Global South that have returned these discards to the Global North raises parallels with the decolonization movement. The key lesson, however, is that "green from the grassroots" is neither liberating nor contributory to resolving the planetary socioecological crisis. If anything, "green from the grassroots" is, overall, exploitative, ineffective, and neocolonial.

If the alleged drivers, consequences, and solutions of "green from the grassroots" are questionable, then alternative responses are needed to the following questions: what are the drivers of plastic pollution in urban centres? Is it lack of access to piped water? Is it poor quality of piped water? Or is it the result of unaffordability of piped water?

The Tragedy of Privatizing Nature

The cause of this waste crisis has been blamed on ignorance and indiscipline of Ivoirian people who do not understand the benefits of sanitation. For the World Bank, it is all the fault of the state: whose inefficient institutional matrices, especially its failure to charge citizens realistic prices (World Bank, 2015, especially pp. 60–62), drive and sustain the problem. That is, inability to recover costs for service delivery is the cause of the problem, together with other state failures (World Bank, 2015, p. 66).

As in much of mainstream environmental economics, new public management, and planning, much of the analyses of causes centre on "state inefficiencies" and inappropriate individual or household behaviour and characteristics. Indeed, everything (see, for example, Sun et al., 2017 for a suggestion on "urban industrial symbiosis," and dos Santos, Svensson, & Padin, 2013, on sustainable business practices) – apart from the monopolization of land and water, the market itself, or the marketization process of land – appears to be at fault. Detailed steps are taken depending on where the analyst places the most emphasis. When Ivoirians are framed as ignorant and undisciplined, the army is used to police waste disposal and discipline those who litter environment (UNEP, 2015). In this sense, making the law in such a way that the cost of polluting the environment would be prohibitive is commonly emphasized (see, for example, Viscusi, Huber, & Bell, 2012; Viscusi, Huber, Bell, & Cecot, 2009).

In turn, the decision rule for what should be done reduces to cost-benefit analyses. The recommendation is usually that if the cost can be recovered through a surcharge on city residents, then state-based curbside (with waste sorted) or other forms of state-based recycling (e.g., single stream programs) are appropriate (see, for example, Aadland & Caplan, 1999, 2006; Bell, Huber, & Viscusi, 2016, 2017; Viscusi, Huber, Bell, & Cecot, 2013; Zen, Noor, & Yusuf, 2014; Zen & Siwar, 2015).

Otherwise, environmental taxes are highly recommended (see, for example, Convery, McDonnell, & Ferreira, 2007), especially if "sold" to the public effectively in messages that emphasize "avoiding a fee" and "paying a tax" (see Muralidharan & Sheehan, 2016, p. 200). Urban public administration texts view the problem in terms of how quickly cities are growing and, hence, focus on canonically improving public administration to deal with the problem. That is evidently the case with Genie Stowers's latest book, *Managing the Sustainable City* (2018).

Yet none of these analyses presents a persuasive explanation. The literacy rate among Ivoirians has been increasing over time, and environmental concerns are taken seriously by Ivoirians (see Djezou, 2014), indeed by most Africans (Njoh, 2014). The rapid urbanization view is even less persuasive. As Beauchemin and Bocquier (2004) have shown, even if we accept that the war in the country pulled people into Abidjan for better services, we should also account for the effect of the war in pushing international migrants and other Ivoirians out of Abidjan to seek safer residences in neighbouring countries and overseas.

When more broadly framed, the most important drivers of the crisis of the commons in Abidjan lie in both neocolonial and neoliberal marketization – the very processes that are prescribed as a solution. Being one of the oldest experiments of marketization, the role of the state, if anything, should be seen as consistent with the marketization doctrine: providing regulation and security for private enterprise. In this sense, state activities are not seen as distinct from market forces, and the market forces are not understood as separate from individual and state activities, but as interdependent and embedded in complex interactions of society, economy, and environment. These processes of marketization are evident in the forces of (1) commodifying water and (2) marketizing waste.

Commodifying Water

As with other West African water governance models (Dagdeviren & Robertson, 2013, 2014), the provision of water has taken the form of privatization of municipal water provision in Abidjan and urban centres in Côte d'Ivoire more generally. Widely regarded as the "French water model" (Komenan, 2010, p. 2) and quite distinct from the Anglophone West African model (Dagdeviren & Robertson, 2013, 2014) in terms of length and the extent of monopoly, urban Côte d'Ivoire has known no other water provider other than the inherited French private company, Société de Distribution d'Eau de Côte d'Ivoire (SODECI), which was founded in 1959 – a year before the Ivoirian independence. SODECI has been in charge of water provision in Abidjan since then, slowly extending its influence to other urban centres from 1973 to date (Traore, 2000). Through its profit-oriented price-setting practices, this monopolistic water provision system has led to recurrent increases in the cost of household water consumption. In response, residents tend to substitute or complement their water needs by purchasing plastic-packaged water (Appessika, 2003; Johnstone & Wood, 1999; Obrist et al., 2006).

While private water supply existed in Abidjan for decades, plastic pollution is a relatively recent phenomenon. Hence, it can be difficult to see an immediate relationship between private supply and plastic crisis. Looking at the history and politics of the plastic-packaged water market makes the relationships clearer. The market for plastic packaging did not spring to being until a relentless campaign against traditional practices of sharing water called it a public health disaster. The solution to the public health crisis, advocates argued, was for individuals to buy packaged water in plastic, store, and use it. This was an intriguing case – combining convenience, good health, and efficiency – to make plastic-packaged water look fundamental for the modern African life. Writing under the headline "The Age of Plastics," Stevens (2002), forcefully argues that "plastics are so clearly useful that it is foolish not to afford them major respect ... Their low cost has undoubtedly had life-saving consequences, as in drought-prone areas of Africa where lightweight plastic-packaged water pails, at times the most important family possession, have replaced clay and stone containers, making it possible to bring in water from even distant wells in times of severe water shortage" (p. 3).

Both globally and regionally (see chapter 7), forces outside Côte d'Ivoire were cumulatively supporting the transformation. Whether in the form of the changes in global governance or the rise of particular ideas such as neoliberalism, the mood complemented the transformation. Qualitatively, the mood in the 1960s was quite different from the mood in the neoliberal era of the 1980s. For Côte d'Ivoire, however, the marketization of the latter years was patterned after the colonial control of resources by France. This specific "French urbanism" (Njoh, 2016) itself appears similar to what was happening elsewhere. For example, in West Africa, neighbouring countries such as Ghana were speedily catching the plastic fever, much like elsewhere in Africa, such as Kenya (Njeru, 2006) and Uganda (Balcom & Carey, 2020). Yet because France continues to exert a much stronger influence over Côte d'Ivoire, the story of "free" markets takes a distinctive form in Abidjan.

In Abidjan, indeed in Côte d'Ivoire more generally, private entities soon entered the market, becoming more prominent in the first few years into the millennium, around 2005, according to the researchers interviewed for this book. Since then, medium-sized Lebanese owned (but locally based) companies have become more prominent in the production of plastic-packaged water be it in sachet or bottled form. Most of the companies' workers are Ivoirians who work under difficult conditions. Bigger plastic-packaged water companies are European (French and Belgian, typically) in origin and in the ownership hierarchy. The "traditional" plastic bagged water dominated in households, but it is less and less in evidence these days, except in poorer areas, such as Adjamé and its market. The corporate-produced plastic-packaged water includes Awa, Céleste, and Olgane in plastic bottles, and Pureté in sachet form. Supported by aggressive marketing on national television and radio, plastic-packaged water companies and the state made plastic-packaged water a star, superior in every sense to sharing water. Across the continent, buying water became widely perceived as better than sharing water.

The demand for packaged water soon increased in response to these aggressive advertisement and public health campaigns. Research experts in Abidjan with whom the issues were discussed stated that individuals purchase packaged water because of the growing high-class cohorts purchasing it in Abidjan, the seeming high quality of packaged water, the convenience, and the ability to store it. The demand for water had an important effect: a fall in price and a rise in profit levels, which, in turn, attracted more entrepreneurs into the market.

Much like in the advanced capitalist societies in the Global North, where the sale and purchase of plastic-packaged water are widespread, even though tap water is deemed clean, through aggressive advertisement, some under the guises of public health campaigns, the "culture" of drinking plastic-packaged water became rooted in the psyche of Ivoirians. Here, the "revealed preferences" of individuals were, in fact, moulded over a long period of aggressive and competitive advertisement that extolled the virtues of the product. Supported by state laws that have evolved under international supervision, there has been a boom in the demand for plastic waste – a process that, as this account shows, was "embedded" in a wider social context (Polanyi, 1944/2001). The economy of plastic-packaged water, then, is as Polanyi (1957) regarded markets, "an instituted process." Rather than individuals autonomously revealing their preferences by buying plastic-packaged water, the market is clearly a social institution. Similarly, it looks like markets are not "natural," as the new institutional economics literature reviewed suggests.

Over time, this constructed market for plastic-packaged water also became class based. Depending on social class, a person buys either bottled water or sachet water – the upper classes going for bottled and the rest for sachet water or, as some prefer to call it "Adjamé water," after the large informal market of Adjamé where sachet water is more in evidence. These changes reflect price differentials. Taking Awa as an example, on average the large size bottle (1.5 L) sells for US$0.82, the next in size sells for US$0.49, while sachets sell for US$0.08. These markets were articulated and supported by one logic: selling water as a superior way of managing the commons. This class dimension to the constructed demand for plastic-packaged water is similar to that in other cities in West Africa (Adams et al., 2020; Obeng-Odoom, 2014d; Stoler, 2012; Stoler et al., 2015). The effects of the boom in plastic-packaged water has been a related boom in plastic waste in Abidjan.

Marketizing Waste

The Ivoirian state relies on corporate waste collection companies to address this problem. Two of these companies are Agrouté and Société Abidjan Salubrité. They work in Cocody and Port Bouet, respectively. A few of the private companies are, however, more prominent. As Sandra Brechbühl (2011) notes, the sanitation situation in Abidjan is decided by four private sanitation companies – LDS, Ciprom, Cleanbor, and Intercor. Even if, in principle, they are all regulated by and answerable to the national regulator, Agence National de Salubrité Urbain de la Côte d'Ivoire (ANASUR), they are private.

Unlike SODECI, the "monopoly" of these companies is partial and de facto but much like SODECI, these private entities have not succeeded in their mandate. In 2009, they could only manage 46.1 per cent of the 893,330 tons of total solid waste generated in Abidjan (Brechbühl, 2011), leaving the rest to be scavenged by vultures, to the pleasure of the wind, or to running water that carries the garbage into the sea, into the lagoon, or into other water bodies. The rest of the waste waits to be collected by precarious labour exploited by private industries. Current total waste levels in Abidjan have jumped to one million tons per year without any corresponding increase in the coverage of the private sector (Andrianisa et al., 2016), so the gap has to be filled somehow.

The use of green technology, especially green plastics, has been proposed as a solution (see, for example, Stevens, 2002). Advocates believe that biodegradable plastics will not stay in the environment for as long. In 2014, the Ivoirian government introduced a law supporting the use of green technology. The challenge, however, is how to determine which plastics are biodegradable in a society with extensive informal economies. More fundamentally, major plastic-related challenges will not go away even with a successful implementation of biodegradables. Exploitation and the non-sorting of waste are two examples. A national implementation of a green plastics program cannot provide an effective solution.

It is possible to use the market in a different way. Here, vendors of water will get a certain amount of money back from the private company, if they return a certain amount of plastic containers. This model could also give incentive to the vendor to collect bags together (for some free "pure water" to be used or sold), while to users, the model could provide an incentive (free "pure water" for use) to return used plastic to the vendor. Vendors and users may also be given some money for their respective roles. In all these cases, the incentives must be sufficiently large to encourage them to collect the waste. However, no matter how large the incentive, wealthier people may not be sufficiently incentivized to make lifestyle changes. Elsewhere in West Africa (see Obeng-Odoom, 2014d), variations of this market model have been unsuccessful for the three reasons given earlier: first, the rate of waste generation has been much faster than the rate of waste collection. So the waste problem remains, worsening over the years. Second there are serious health dangers for the waste pickers. Third, it is unclear but potentially risky to depend on the waste returned to human society without careful health assessment of how clean this "recycled" waste is.

Conclusion

Cities face pandemic ecological crises that threaten their nature and future as the common meeting point of humans and other living and non-living organisms. The Conventional Wisdom that this crisis arises from the "tragedy of the commons" was presumably questioned by Elinor Ostrom. However, what Ostrom questioned was that neither market fixes, technology, nor privatization is the "only" way. She proposed that all these could be part of the solution. In addition, Ostrom proposed and defended support for, and governance of, urban CPRs, such as informal communities, which, according to her, arise based on a rational rejection of Leviathan, state-centric urban governance and provide a potent additional grassroots approach to addressing the ecological crises in cities. These arguments are, therefore, part of the Conventional Wisdom. As discussed in this chapter, the Conventional Wisdom overlaps with the Western Left Consensus on the vitality and agency of informality.

This chapter has called these claims into question, however. Drawing on, among other factors, first-hand experiences of cities in Africa, this challenge necessitates fundamental rethinking of the orthodoxy. Informal urban common pools or communities remain widespread but not so much because of a rational decision to self-govern. Rather, they grow from oppression and work under suffocating conditions. Privatization has its place in society, but markets have not deterred polluting behaviour; they have augmented it, a point developed further in chapter 6. If there is a tragedy, it arises from the privatization of nature, which, as the case of Abidjan shows, generates substantial waste.

The contribution of waste pickers is a good example of what Ostrom (2012a) calls "green from the grassroots" in the urban commons literature. However, while celebrating hyper-agency, this polycentricity ignores the structural limitations that green waste pickers recurrently face. Clearly, the idea that, through market instruments and individual responsibility, the urban socioecological crises could be addressed overlooks the central structural challenge: the monopolization of land seemingly to address the "tragedy of the commons." Widespread technological adoption might be one way of addressing this problem. Indeed, in his book *The Urban Commons: How Data and Technology Can Rebuild Our Communities* (2018), Daniel O'Brien argues the case about how technology provides the ultimate solution to the "tragedy of the commons." In a complex world that is inherently uncertain, O'Brien places substantial faith in the healing power and certainty of big technology. Such claims raise important questions. For example, can the wave of technological uptake be sustained? What about the *relative* benefits of this technological breakthrough: do they facilitate inclusive development at the local and global scales? These are questions that I address in the next chapter.

Chapter Five

Technology

Introduction

The question of sustaining technological momentum is complex. On the one hand, the Conventional Wisdom claims that strict patent regimes are imperative. This idea, that without enclosure, innovators would not have the incentives to invest in technological innovation and advancement, was first discussed in neoclassical economics by Richard Nelson and Kenneth Arrow (see, for a detailed discussion, Allen & Potts, 2016; Potts, 2018). In the commons debates the idea arises from a concern with the "tragedy of the commons" (see Chen & Puttitanun, 2005; Lerner, 2009, for an overview of the economics of patenting and innovation, as well as its declining influence) into which can be shaded the problem of market failure. So, as innovators cannot privately appropriate the fruits of their labour – because knowledge is a commons – they are unlikely to have the incentives to invest in innovation. If they do, without adequate pricing, the innovation is likely to be overused and abused. This problem of underproduction and overuse can be solved only through the introduction of market instruments such as patents to privatize the benefits of innovation and make its overuse and abuse costly (see, for a detailed discussion, Allen & Potts, 2016; Potts, 2018).

On the other hand, the Western Left Consensus sees far more potential for collectivizing the use of technology or for making technology a commons. The case for greater commoning is complex, but two themes are particularly important. First, much of the technology in existence today arose from the innovation commons, that is, from a loose group of social practices that were neither in the domain of the state nor in the context of the market (firms). If so, mainstream innovation economics, which is based on state/market relationships and how they shape innovation policy, ought to be fundamentally problematized. Indeed, more, rather than less, commoning is needed to sustain the technological advancement that characterizes the present era (Allen & Potts, 2016; Potts, 2018). A second theme more directly emphasizes the political role of the technological commons to undermine capitalism, to provide new paths for a commons-based system of production and exchange, and to facilitate the transition into a more socially inclusive world (Euler, 2016; Niman, 2011; Papadimitropoulos, 2018; Wright, 2010, pp. 194–203). The open source movement, free software advocacy groups, the creative commons advocacy, the copyleft actvists, and Wikicommons groups are all examples of, or advocates for, commoning technology, which can easily become both imperial and imperializing, as the example of Facebook shows (for a briefing, see "Briefing," 2016; Bruncevic, 2017, pp. 187–205). Over a much longer period, writers as diverse as Thorstein Veblen, Joseph Schumpeter, and Lewis Mumford all recognized the potential of technology and with varying degrees of optimism accepted it. However, all put the case for commoning and developing its ethical foundations. Lewis Mumford considered that, by their nature, cities could become fruitful locations for developing such technological commons (for a detailed historical account, see Jamison, 1998). More recent research appears to take this view, too, as the work of Alessandro Aurigi and Nancy Odendaal (2020) shows.

The second question, centred on the distribution of technological effects, is even more vexing. Consider the Sustainable Development Goals (SDGs). They are usually held out as humanity's agreed hope for an inclusive, stable, and sustainable future. Yet these goals represent the Conventional Wisdom that I have been discussing in previous chapters. Anchored on growthcentricism, they rely on technology to frame this growthist vision (see, for example, Goal 8, and many other studies such as Gunderson, 2018; Hickel, 2019; Nagendra, 2018; Robra & Heikkurinen, 2019). Goal 7 and Goal 9, for example, illustrate the point. Other examples are responsible consumption (Goal 12), sustainable cities (Goal 11), and reduced inequalities (Goal 10). Urbanization itself is regarded as a problem, so is increasing global population (Goal 15). Technology is viewed as a panacea, with the Charter Cities Institute becoming the latest (since 2017) to add its voice and weight to the calls of others such as the UN. However, the systematic evidence (see, for example, Castells, 1977, 1989, 2010) about these supposed linkages between technology and sustainability have often been weak. Similarly, the comparability of the evidence is questionable because of wide variations in their underlying methodologies (Shinwell & Cohen, 2020). Yet, in the light of the many iconic smart cities developed in Masdar City in UAE and Songdo in South Korea, Silicon Savannah in Nairobi,

Figure 5.1. The Radical Alternative on the Technological Commons.

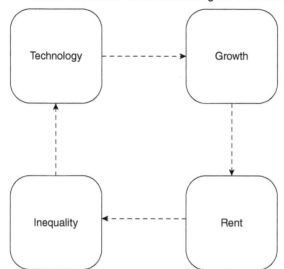

Sources: Adapted from George, 1879/2006, 1883/1966, 1898/1992, 1891, 1892/1981, 1886/1991.

and Yabacon Valley in Lagos (Grant, 2015, Chapter 6; Kitchin, 2015; Manirakiza, 2014; Murphy & Carmody, 2015, 2019; Siba & Sow, 2017), it is important to revisit the criticisms of the SDGs. Among other concerns, the SDGS have been dismissed as top-down, barely reflecting grassroots concerns, and, crucially, inattentive to land. As recently noted by F. V. Noorloos and colleagues (2019), "SDG 11 and the New Urban Agenda have re-energized debates on how to achieve an inclusive and sustainable urban transition in the Global South. However ... 'land' is only discussed in the margins" (p. 856).

The land question is particularly important. It was the central point of discussion in the nineteenth century. The framework in which the land question was discussed is quite comprehensive. Today, that lens remains the most effective in addressing the land question, technology, social change, and sustainable development (see, for example, Daly et al., 1994; Haila, 2016; Niman, 2011; Obeng-Odoom, 2016b; Petrella, 1981).

As schematically shown in Figure 5.1, this framework is centred on technology, growth, rent, and inequality. Technological change – regardless of its specific type – usually increases economic growth.

The mechanisms for doing so are complex, but they include a transformation in the nature of landed property and the more obvious increases in industrial production. The resulting economic growth substantially changes the process of rent formation such that land rent typically increases with economic growth. In turn, landlords are able to extract more and more rent from labour in a process in which the labour share of the surplus generated in the production process *progressively* reduces. So, even though technology makes it increasingly easy to access various parts of the city, workers and citizens more generally have to be housed in slum-like conditions. While finding more work through the expansion of the economy, workers must increasingly become worse off as they pay proportionately more of their wages as rent and live less and less on shrinking effective wages.

Landlords, on the other hand, can increasingly speculate on their land and monopolize the benefits of technology through extracting more and more rent. These landlords control not only where people sleep but also what people have left over to feed themselves and their families. They control the destiny of future generations. They shape the future of the urban economy. They mould how much land, where, and how it could be used for urban development and development more widely. In the hands of the landlords is the power to affect the possibility of famine, to shape economic depression, and to shape inequality in society.

This dynamic towards inequality and stratification, in turn, worsens whenever a new technology is introduced in the city. Regardless of the specific type of technology: when roads are built, when railway lines are constructed, when high-tech shopping centres are raised, when new factories are opened, when new sources of energy are injected into the process of production, that tendency increases. Inequality, then, shapes who is able to appropriate and use technology. Access and control of technology become a product of structures of inequality.

Figure 5.1 suggests that the present global system is inherently unstable and unsustainable. However, the reasons are not that humans have encountered machines, or because of urbanization per se, or even because the size of population is increasing. Rather, it is because our land tenure system is designed in such a way that the benefits of the machines are appropriated by landlords, while the condition of labour deteriorate (Haila, 2016; Kelly, 1981, pp. 299–300).

Addressing these dynamics could require the application of a range of policies. The first is untaxing labour to ensure that it gets a fuller version of its products while promoting a labour-based, anti-monopoly system of production. The second is keeping non-privatized land as commons or, if the land is already privatized, taxing away the rents generated by technology, speculation, population growth, and wider social investment. The third is investing the resulting revenues for socio-ecological purposes. They could include limiting inequality. Breaking up monopolies and supporting small-scale labour-based production are examples of such ways. Doing so could free are examples of such ways. Doing so could free labour from exploitation, nature from domination by capital, and future generations from the tyranny of current humanity. These socioecological ends could be achieved by explicitly preventing the transfer of the concentration of land to only the future generation of landlords. It is these policies that are likely to attenuate the technology ↔ growth ↔ inequality ↔ technology dynamic illustrated in Figure 5.1.

Based on Figure 5.1, three questions can be asked of the SDGs. First, in what ways could technology facilitate economic growth? (See also Chang, 2002; Newman et al., 2016.) Second, what is the relationship between this process of economic growth and inequality? Third, could technology-based growth and its resulting consequences guarantee stability and sustainability, as many SDGs suggest? (See, for example, Goals 8, 9, and 11.)

While many studies, mostly written from the perspective of the Western Left Consensus, try to answer these questions (see, for example, Gunderson, 2018; Hickel, 2019; Nagendra, 2018; Robra & Heikkurinen, 2019), they, much like the SDGs Conventional Wisdom, pay little or no attention to land (Noorloos et al., 2019). This chapter attempts to answer these questions by emphasizing land as discussed in Figure 5.1, the overarching framework of this chapter. One aim is to use existing data to systematically analyse the assumptions about growthism, inequality, and urban ecological modernization, three key characteristics of the SDGs. Another aim is to contemplate possible alternatives to them, as suggested in Figure 5.1. The chapter's claim to originality is not the discovery of new data per se, but rather providing new interpretation on the basis of a re-envisioning of existing data within Figure 5.1.

Based on analysing a range of evidence from the Global South within this framework, it can be argued that inherent to technological change has been the rapid increase in growth, driven largely by technologically mediated speculation. This dynamic, in turn, drives up urban land rent, which is largely fictitious. This uncertainty and instability create corrosive implications for real wages – which would tend to decline over time as more rent or interests are paid. Uneven urban development, produced, among other ways by actions and inactions to enhance speculative rent extraction, also worsens. Combined, these effects could make growth even more fragile and inequality even more structural. Consequently, the emphases on growthism, urban ecological modernization, and technological triumphalism in the SDGs is misplaced. It should follow that the SDGs, indeed every technological, industrial, or employment goal or policy, must pay serious attention to land reform as a crosscutting issue.

The rest of this chapter is divided into three sections, respectively focused on economic growth, inequality, and sustainability.

Economic Growth

As suggested in the SDGs Conventional Wisdom (see, for example, Goals 8 and 11), technological diffusion in the Global South can, indeed, generate growth. The process of growth, however, requires some discussion. The digitization of the urban economy provides one such avenue. Through production-related digitization, financial digitization, and the digitization of urban governance, systematic processes have been triggered to mould growth. In practice, these forms of technological adoption are intertwined (Kitchin, 2015). Production-related digitization relates to a wide range of activities (Datta, 2018; Grant, 2015). Among them is the introduction of technology in factories (including the use of technology in the actual production and in the process of production, such as monitoring factory workers), transportation, retail outlets, and consumption activities. Such digitization also includes the creation of new technology-related jobs such as those offered by Foxconn in China. In informal economies, such digitization plays out in the form of jobs, such as selling mobile phone services and offering phone credit.

There have also been high levels of technological adoption in the financial industry. Banks have increasingly adopted ATMs and mobile money services, including SMS services (Domeher et al., 2014; Matsebula & Yu, 2020). Internationally, banks are better able to do online transactions between the Global South and the Global North and, by using information-communication technologies (ICT), make urban residents aware of the availability of such services. In turn, new technologies have reduced information asymmetries and transaction costs (Asongu & Nwachukwu, 2018). Non-bank financial institutions are also undergoing transformation. The mobile account revolution, which is currently available in 61 per cent of all countries in the Global South, is one example. In East Africa, of every two mobile phone connections, one is a mobile account connection (GSMA, 2014), a trend that has incentivized many traditional banks – notably FNB – to start working with telecommunications networks (Gopaldas, 2016). Mobile telephony is increasingly being used for remittances, for the payment of goods and services such as taxi payments in Kenya, and for monitoring workers, for checking prices, and for setting new ones (Bateman et al., 2019; Duvendack & Mader, 2019; Murphy & Carmody, 2015, 2019).

In a broader shift from just urban governance to e-urban governance, city and national authorities in the South are increasingly emphasizing the importance of digitizing information. Reminiscent of Manuel Castells's famous dictum that "first, the core new technologies are *focused on information processing*" (Castells, 1989, p. 13, italics in original), the state uses geotechnology and the digitization of land information in various land administration projects in the Global South. The state also uses its power of guaranteeing contracts to change the ways in which land is accessed, acquired, and exchanged. Seen as part of the ICT4AD or ICT for accelerated development, the state has been a key driver of digitizing urban land information (Karikari, 2006; Karikari et al., 2003) to create land supermarkets (Obeng-Odoom, 2020). A major part of the land titling programs rolled out in the Global South, this attempt to convert land into commodities by creating large land information systems includes the use of geotechnologies such as GIS to digitize information and ensure the preparation of land titles and land registers can be accessed globally. Real estate agents are increasingly using ICT and mobile telephony to help in this process of transformation (Akaabre et al., 2018). As land information and transactions become increasingly available on the internet, itself more prevalent in cities, the process of digitization intensifies. In India, about £11 million was allocated in 2014 alone to make governance transparent and accountable through the application of technologies in a process, which the Indian urbanist Ayona Datta (2018, p. 410) argues is changing the vocabulary of "cityzens" to "netizens." So, urban space is important, but it has become even more important with the creation of networks of ICT.

How have these changes contributed to growth? Analysing technologically driven urban economic growth can be difficult in the Global South where such data are not systematically collected. One way to mitigate the challenge is to consider intra-firm and inter-firm clusters and household effects of technological adoption, as well as the ramifications of technology for transactions at these individual, household, and firm scales. Likewise, considering intra-city, inter-city, intra-national, international, and global scales is useful (Murphy & Carmody, 2015, 2019). Across these scales, considering the role of technology in shaping transaction costs for local import and export transactions can be a useful point of departure.

Technology has enabled urban economic growth, generated from within and between cities. Within cities in the Global South, small and medium-scale enterprises (SMEs) have experienced "incremental improvements to efficiency and productivity" (Murphy & Carmody,

2015, p. xxi, 2019). Between SMEs, especially within their clusters, there is much sharing of information and technology, as well as labour pooling, which increases efficiencies and productivity. But localization economies are not the only drivers of urban economic growth. There are also urbanization economies, advantages that are much wider than those produced by firms and are contributed by a vast array of actors, as has been established for Chinese cities (Chen et al., 2016).

Improved technology has also better positioned cities to attract macro-economic gains. These were always possible, of course, but they are even better now. Nation-wide gains in growth are increasingly being captured and concentrated in cities. As more of the national population is clustered in cities, more technologies converge in cities, and more production and consumption take place in cities. Policies that are meant for nation-states as a whole become concentrated in cities, too. Consider the case of monetary and fiscal policies. They are set for entire nations, but cities become important points where they are shaped. As they now possess the technological tools required by banks to operate, cities are increasingly welcoming new banks. Take Accra. It has only 13 per cent of the population of Ghana, but it is the seat of 35 per cent of bank branches in the country (Obeng-Odoom, 2011a) and the headquarters of most of the major banks in the country. Interest rates are shaped by the vibrant bank activities, including generating loans – as are government expenditure patterns. For example, more than 50 per cent of all national oil revenue went into the production of urban roads in Ghana between 2011 and 2013 (Obeng-Odoom, 2015e, 2020). The spread of roads has substantially facilitated the boom in telecommunications. This nexus has been established (Okyere et al., 2018) for cities and regions where, statutorily, the construction of roads must create way leaves for telecommunication lines.

This micro-macro intermingling is further shaped by global forces, which, again, work to improve urban economic growth. African urban economies are also becoming even more strongly integrated within global urban economies. Analyses of African airline connectivity, the volume of airline passengers, and the direction of flights from Africa to elsewhere and from elsewhere to cities in Africa show dramatic increases (Otiso et al., 2011). Between 2001 and 2009, for example, the volume of increase in airline passengers ranged from 34 per cent in Harare, Zimbabwe, to over 1000 per cent for Marrakech in Morocco and Lagos. So the roads in cities in Africa become longer and wider to accommodate more cars and motor bicycles fuelled by transnational resource corporations such as Total and Shell (Obeng-Odoom, 2018). Also, the airports in cities in Africa have become particularly busy spaces, especially in Johannesburg, Cairo, Cape Town, Nairobi, Durban, and Casablanca.

Much like for cities elsewhere, the technologically driven urbanization in the Global South has contributed substantially to growth. For example, it has been estimated that, in Africa, 10 per cent mobile phone penetration increases GDP by 1.4 per cent (Grant, 2015, p. 148). São Paulo, with only about 10 per cent of the Brazilian population, contributes more than 40 per cent of the national GDP of Brazil. On only 0.1 per cent of land, Mexico City contributed 30 per cent of the national GDP of Mexico, while Riyadh, on less than 1 per cent of land, contributed about 20 per cent of the GDP of Saudi Arabia. The case of Africa is even more intriguing. Cairo occupies only 1.5 per cent of land in Egypt, but it contributes more than half of the country's GDP (World Bank, 2009). Luanda, on only 0.2 per cent of land in Angola, produced about 30 per cent of Angola's GDP. Nairobi, Casablanca, and Lagos produced about 20 per cent of the GDP of Kenya, Morocco, and Nigeria on less than 1 per cent of land. These figures are problematic. Aside from questions of estimating cause and effect, there are also concerns about the quality of the GDP figures (Jerven, 2015), questions about measurement of informal urban economic activities (Grant, 2015, p. 148), and issues related to new forms of fossil-driven urban economic growth (Obeng-Odoom, 2014b, 2020). However, even if imprecise, this growth record is indicative of the contribution of technology for urban economic growth.

Inequality

The distribution of this growth is a major issue. Inequality in cities in the Global South, often analysed superficially in the Western Left Consensus, is far more widespread and definitely more complex than often presumed (Obeng-Odoom, 2020). Its most visible manifestation is the spread of gated housing pockets planted in oceans of decrepit housing and slums that receive no or low municipal services. Economic inequality in the form of unequal income and wealth distribution is less visible but widespread nevertheless. As Table 5.1 shows, urban economic inequality is on a continuum from the least to the most unequal.

That cities in Africa are the most unequal in the world (UN-HABITAT, 2008, p. xiii) is serious enough but, as Table 5.1 shows, inequality is more widespread. Inequality between labour, capital, and landlords is one form, but it manifests in different ways across cities in the Global South, where capital cities surge ahead of secondary cities, and regional capitals take on a new form of importance compared to provincial towns (Obeng-Odoom, 2013a, 2020). Inequality between cities in the South and cities in the North is another form of inequality occasioned by technological dependency (Murphy & Carmody, 2015, 2019).

Table 5.1. Urban Economic Inequality in Selected Countries in the Global South

Low Inequality Group (Gini Coefficient of Less Than 0.299)	Relatively Low Inequality Group (Gini Coefficient 0.300 to 0.3999)	Relatively High Inequality Group (Gini Coefficient 0.400 to 0.449)	High Inequality Group (Gini Coefficient 0.450 to 0.499)	Very High Inequality Group (Gini Coefficient 0.500 to 0.599)	Extremely High Inequality Group (Gini Coefficient 0.600 to 1.000)
Belarus	China	Cameroon	Philippines	Argentina	Namibia
Romania	Poland	Uganda	El Salvador	Brazil	Zambia
Bulgaria	Lithuania	Côte d'Ivoire	Peru	Chile	South Africa
Armenia	Algeria	Vietnam	Venezuela	Colombia	
Serbia	Georgia	Nepal	Bolivia	Ecuador	
Hungary	Tajikistan	Malaysia	Mexico	Zimbabwe	

Source: Adapted from UN-HABITAT, 2008, p. 63.

Land speculation leads to increases in rent and in land value, which accrue to landlords, many of whom are absentee owners who speculate on their land to increase rents. It is, thus, through a process of cumulative change rather than a single change that rents arise and rise.

As suggested in Figure 5.2, the speculation on land values is a major part of the process across scales (national, international, and global). Speculation shapes both the form of production and its spatial expression. Coupled with this dynamic, technology enables a new panopticon. Employers can better monitor labour activities and better demand more time from labour – without necessarily improving conditions of employment. Thus, even if the working day is not lengthened, its intensity is increased through stricter monitoring (see Murphy & Carmody, 2015, 2019) and more opportunities for "nibbling and cribbling at mealtimes" (Marx, 1867/1990, p. 352). This phrase, used in *Capital*, Volume 1, to describe what today would be the equivalent of work demanded by bosses during workers' breaks, after work, and on weekends through phone calls and emails, has become even more relevant. Such has been the labour experience in tech companies such as Foxconn in China. Working for Apple and HP transnational companies as an electronics contract manufacturer, this technology giant is well known to use exploitative practices to reduce costs and enhance profits for capital both internally (in China) and internationally (for the United States) (for details, see Bieler & Lee, 2017; Lüthje & Butollo, 2017). SMEs face other pressures. As an example, international customers are not entirely trusting of SMEs in tourism. Accordingly, a survival strategy has been for them to become

Figure 5.2. The Creation of Urban Economic Inequality.

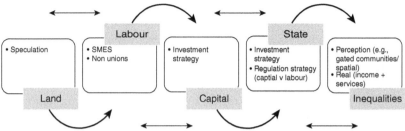

Source: Obeng-Odoom, 2017a.

subcontractors of bigger networks, such as Trip Advisors and Hotels. com. The implications for global income distribution are wide-ranging (Murphy & Carmody, 2015, 2019). Micro businesses and weaker SMEs incapable of accommodating the transaction costs of working with intermediaries have had to *downgrade* by selling cheap, thus cheapening and exploiting labour (Murphy & Carmody, 2015, 2019).

It is not only these dynamics of labour that contribute to the production of inequality, enabled by mobile technology. The logic of investment strategy of capital under the new technological age complicates these experiences. In contrast to urban economics location theories, which suggest that firms respond to location decisions (for a discussion, see Obeng-Odoom, 2018) in partnerships with the state, technological giants such as HP, IBM, and Siemens also shape their location (Datta, 2018). Thus, Foxconn, a tech company operating in China, has effectively determined where to locate by dictating its terms to the provincial state, effectively concentrating its operations in Shenzhen. The company has been lured to extend its factories to other cities such as Chongqing, after the state agreed to upgrade its airport, and Zhengzhou, where the urban government provided more high-tech import facilities to reduce the company's transaction costs, enhance its efficiency, and, hence, improve its profits. Similar comments apply to Huawei located in Shenzhen (Zhang, 2015, pp. 121–123). Giant tech companies are, therefore, becoming new municipal governments.

Indeed, the reliance on liking Facebook posts about policy proposals for cities, evaluating urban policy by relying on YouTube viewings, and tweeting of urban policy proposals exemplify the dominance (Datta, 2018). State strategy has reinforced the spatial concentration of opportunities and the effect of technological parks has been to concentrate rather than to spread economic development. Thus, colonial urbanism of spatial segregation has continued and even intensified through "circular and cumulative causation" (Myrdal, 1944). Gated communities produced for TNCs are emerging in cities such as Sekondi-Takoradi (Obeng-Odoom, 2014b), where urban farmers have been displaced to create such communities in a process that has created structural incentives for land speculation. So, developing a land-based explanation for the rise of gated housing has become an important pillar for urban economic transformation (Ehwi et al., 2019).

In this process, both landlord and capitalist shares in the fruits of growth increase, while the share of labour declines. Or the share of profits soars, while that of wages merely increases. The reasons for this contrasting experience can be explained in a number ways. One is that labour recurrently suffers from eviction, exploitation, and precarious downgrading without much opportunity for unionization and the pursuit of social and economic rights (Flanagan & Stilwell, 2018; Obeng-Odoom, 2017b, 2020).

These problems generate and maintain the social foundations of instability. What requires further analysis is how such instability structures socioecological sustainability. Although tentative responses can be gleaned analytically from Figure 5.1, the overarching framework for this chapter, this logical plausibility can usefully be complemented by ascertaining its congruence with experiences in the Global South. Sustainability provides a useful point of departure.

Sustainability

Systematizing the evidence must entail an analysis of how the Conventional Wisdom envisages trends. Both the SDGs and many urban economics growth models predict that the inequality and instability analysed in this chapter's previous two sections would decline, as economies continue to grow (see, for example, Gunderson, 2018; Hickel, 2019; Nagendra, 2018; Robra & Heikkurinen, 2019). Target 1 of Goal 10 seeks to "progressively achieve and sustain income growth of the bottom 40 per cent of the population at a rate higher than the national average." This point is echoed in the recent contribution of many economists, including Antonio R. Andrés and Anutechia Asongu, who forcefully defend this ecological modernization and sustainable economic growth approach in their contribution to *Technology in Society* (see Andrés & Asongu, 2019).

The empirical evidence, however, is far more complex. Inequality has continued to soar, although GDP has been increasing, as is the case in Jakarta (Thynell, 2018) and Africa more widely (Fosu & Gafa, 2020). Indeed, inequality has choked off poverty reduction in cities such as Accra (Obeng-Odoom, 2017a), while in Ho Chi Minh City, inequality

Figure 5.3. Technological Mediation, Growth, Rent, and Inequality.

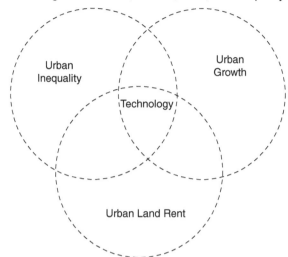

has created a middle-income trap (Gore, 2017). These are not mere correlations (see, for example, Fosu & Gafa, 2020).

As illustrated in Figure 5.3, the contradictions in technologically driven
growth are inherent. The sheer magnitude of the amount of evidence that
demonstrates this relationship (see, for example, Ndiaye, 2020; Piketty,
2019) is indicative of causation, not mere correlation (Wilkinson & Pickett, 2019). Additional examples illustrate the point.

Urban growth undermines real economic growth in the long term. In
the case of China, where local banks are increasingly providing credit
for real estate development in cities, it has recently been shown "that
credit expansion held back consumption growth by claiming resources
for investment in gross capital formation and net exports" (Zhang &
Bezemer, 2016, p. 613). This problem is likely to worsen, as is well
known of instabilities inherent in speculative finance (see, for example,
Minsky, 1992; Veblen, 1923/2009).

India's urban economic growth, which has given it the status of a
so-called rising power, has been made possible through technologically mediated exploitation of darker races. The appropriation of the
land of Indigenous people by absentee owners who, thanks to technology, can speculate even more on Indian land from across the world is
another example (Talukdar, 2017). The continuing dominance of TNCs
in driving uneven urban development has also exposed as myth the
idea, strongly defended by Matthew Khan (2010), that fossil-heavy

urban pollution can be addressed by new technologies (Obeng-Odoom, 2016b, 2020). New Delhi and Bengaluru, in particular, recurrently faces the problem of urban pollution, while residents in cities in Africa now inhale air that is 100 times as sulphurous as what urban residents inhale in Europe (Nagendra, 2019; Obeng-Odoom, 2018, 2020).

This new international division of labour, typified by increasing monopoly of capital, "wage theft," and "rent theft," further deepens global inequalities. Such inequalities have serious negative consequences for global peace, damning consequences for prosperity, and devastating consequences for the environment (for a detailed discussion, see Stilwell, 2017, 2019b). When nature ceases to be the source of wealth and nourishment for humanity and instead becomes the private property of a few lords for whom only speculation and more rent are important, the problem of inequality becomes particularly serious.

Inequality has damaging effects on who benefits from technological innovations. As Figure 5.3 suggests, when finance is concentrated in the hands of the few, only they benefit from the technological innovations that now define the finance industry in the Global South (Gwama, 2014; Matsebula & Yu, 2020). Thus, at the local level, inequality undermines financial development, creating economic fragility that could collapse growth. Similar dynamics apply at the global scale. Global urban inequalities have not inhibited the diffusion of technological advance, but it has limited the creation and control of technology in the Global South (Murphy & Carmody, 2015, 2019). With the offshoring of inferior activities made possible by technology (Robert-Nicoud, 2008), the Global South is assigned dependent industrialization that is characterized, among others ways, by labour precarity, which, in an environment of increasing rent, makes the conditions of labour particularly dire. Indeed, inequality structures technological change, shaping who uses technology, which ones, and for what purposes. These inequalities are not only intragenerational but could be intergenerational, too. The children of landlords inherit their world while the children of the rest inherent their disadvantage.

For cities, the price of inequality is particularly high because inequality undermines urban sustainability – whether conceptualized as economic, social, or ecological (Stilwell, 2000, 2019b). If conceptualized as economic sustainability, that is, the avoidance of waste, the production of maximum goods and services with limited resources, and the reproducibility of goods and services, inequality is anathema. Economic sustainability is undermined by increasing rent, and the declining labour share contributes to a shrinking urban economy and fragile growth. Widespread waste and downgrading arising from hyper-competition can also be ascribed to increasing inequality. Urban sustainability as social sustainability is much broader. Such spatial sustainability entails the recognition, maintenance, and nourishing of social practices and systems, as well as the institutions on which they are based. Although wider, socio-spatial sustainability is similarly undermined, albeit through different inequality mechanisms, such as social conflicts (often racialized/ethnicized). Centred on land, such as the coup in Madagascar, and the social control of an entire continent by absentee owners, this dynamic is disturbing. Urban ecological sustainability is a much wider form. It entails a recognition that inequality be limited.

Yet urban ecological sustainability, too, is under threat. Urban pollution from increased car "waste" imports and "unequal exchange" is one example. Top-grade oil is appropriated from the continent and used in Europe, from where the dirty oil is exported to Africa. Is it any wonder that urban biodiversity is declining on the continent (Dawson, 2016; Obeng-Odoom, 2018, 2021)?

The SDGs are admirable as a vision. SDGs 10 and 11 focus on reducing inequalities and promoting egalitarian cities. More inclusive urban development is desirable (for an extensive discussion, see Stilwell, 1992, 2017, 2019b), not only because such problems could be attenuated, but also because the empirical evidence shows that more egalitarian societies are, even on a narrow economistic efficiency criterion, more efficient. Scarce resources are better channelled into more productive sources, such as employment creation, rather than, say, security from perceived attacks by the poor who live outside urban gated communities and in the townships and slums. On a broader level, more inclusive societies are happier, safer, more peaceful, and healthier. Politically, more inclusive societies are also more responsive to a broader array of interests that essentially controls the state, the economy, and the wider society. Thus, this aspiration is a better vision for the future of our cities.

The trouble is that the SDGs are placed on shaky foundations. Whether in terms of technology, growth, or inequality, the SDGs are contradictory. As illustrated in this chapter (see Figure 5.1), without considering land, how technology transforms land rent and inequality, and their ramifications for society, economy, and environment, the SDGs are highly problematic. Not only do they fail to call into question growthcentrism, as previous studies have shown (e.g., Gunderson, 2018; Hickel, 2019; Nagendra, 2018; Robra & Heikkurinen, 2019), but they also promote more and more growth. Going by the rule of 72, a growth rate of 7 per cent, which is the target of Goal 8, would double the world economy in only 10 years' time, that is, around the end of the SDGs timeframe of 2030. With technologically driven growth based on the transformation of land (see Figure 5.1) being so central to the sustainability crises, the evidence analysed in this chapter shows that the SDGs undermine sustainability. Worse, they provide instead the mechanism for growth, but only with increasing inequality, instability, and unsustainability. The future needs to be radically different.

Conclusion: Towards the Future

If the SDGs represent the Conventional Wisdom on the technological commons, with its many problems, as I have demonstrated in the previous sections, is the Western Left Consensus more compelling? Much like Karl Marx who "refused to either accept or reject technology, but rather dissect it" (Heideman, 2015), modern political economists are critical of technology without rejecting it (Stilwell, 2006, pp. 332–341). They propose, instead, what can be called a Western Left Consensus, that is, appropriate technology for a new economy. Much like what is now clearly a cliché, the Lucas Plan of 1976, for which workers of Lucas Aerospace, facing the threat of redundancy, took over technology to show how it could be socially useful, this Western Left Consensus is centred on the socialization and worker control of technology for social purposes (Palmer, 2017, p. 5). Today, social scientists and others tend to demand appropriate technology or endogenizing technology so that patterns of dependency can be broken. In popular practice, many former tech developers, disenchanted by the corporate control of technology by Google, Facebook, Samsung, and Microsoft – whose hidden costs work to the advantage of technology TNCs – have formed their own open technology movements, including Boot Camp Rebels (Ram, 2017). Indeed, currently there are about 1883 organizations in the EU alone developing what they call "tech for good" (Ram, 2017, p. 8). This trend is potentially radical.

It is even more radical than the vision of Sir Tim Berners-Lee, the British inventor who developed the worldwide web. For Sir Berners-Lee, his invention was intended to "allow a place to be found for any information or reference which one felt was important" and that "you can't propose that something be a universal space and at the same time keep control of it" (Ram, 2017, p. 8). For this current movement, it is workers and citizens generally who must own these technologies. In his time, Thorstein Veblen preferred engineers to do so because they were particularly better placed to control technology. For Veblen, doing so would enable society to benefit from the use of technology for real production, especially if the engineers were socially conscious (for a discussion, see Jamison, 1998; Veblen, 1923/2009).

At the urban and regional level, similar aspirations have been expressed. David Harvey, the most prominent urban political economist today, considers them to be "the Future of the Commons." According to Harvey (2011), the "Swedish Meidner plan proposed in the late 1960s ... ought to be reconsidered." We need a "tax on corporate profits, in return for wage restraint on the part of unions, ... placed in a worker-controlled fund that would invest in and eventually buy out the corporation, thus bringing it under the common control of the associated laborers" (Harvey, 2011, p.105). Much broader proposals – of which the "one million climate jobs" campaign is the most visible – emphasize using green technologies to create evenly spread green jobs and structurally inclusive urban and regional development (Campaign against Climate Change, 2014; Stilwell & Primrose, 2010). Where they exist or could be created, such jobs have the potential to ensure that technology aids inclusive urban development and agendas set by local urban residents (McFarlane & Söderström, 2017).

In this regard, the activities of several NGOs supporting low-tech, small-scale technologies in use in the Global South have been commended. So are the activities of ordinary urban citizens (Acey & Culhane, 2013; Nagendra, 2019). Such technologies include simple water heating devices and biogas technologies that convert food waste into fuel, a project locally produced in Cairo and currently supported by C³ITIES (Acey & Culhane, 2013). It is important, however, to explicitly link these appropriate technologies and their multi-level tiers of support to wider macroeconomic policy (Spies-Butcher & Stilwell, 2009; Stilwell & Primrose, 2010). Doing so could address the twin problems of urban environmental crisis and urban economic recession or simply low urban economic dynamism.

Such changes, proponents argue, would restructure the urban economy in ways that could lead to some job losses (e.g., by shifting away from some types of polluting technologies to embracing sustainable technologies and small-scale low technologies) but new jobs would also be created. These "green jobs," as they are called, would have secure tenure, shorter working days, and improved working conditions. There would still be growth – not growth for the sake of profit but for a better, greener, more inclusive society in which growth is worker-driven and controlled. In this sense, the critique of the triumphalist story also includes democratizing work and demanding not just economic but also social and environmental justice.

The proposals for the future city look quite comprehensive. The ongoing discussion is not simply a dualistic debate of technology or no technology. It is much more complex. It includes questions about what type of technology, how it is created, for what purpose, under what conditions, who owns it, how it is accessed, at what cost, and what the role of technology is in urban economic development and development more widely.

These discussions and debates, generated by proposals for appropriate technology and green jobs, are important. The explicit interest in using ICT for development is important, too. Without reforming urban land tenure systems in a way that would ensure that socially created rents are socially appropriated, growth could continue, this time driven by workers, but much of it could be fictitious, inequality could worsen, and nature ultimately could be lost, if absentee ownership continues and the lords remain in control of the land under our feet. From this perspective, Paul Romer's (2010, 2018) idea of "charter cities," better classified as the Conventional Wisdom, is highly problematic, partly because (1) it is neocolonial in insisting that new cities governed by former colonizers be exported to Africa, (2) its claim that Hong Kong has been successful because of private land ownership is empirically flawed (land in these city-states is publicly owned; see Haila, 2000, 2016), and, particularly, (3) charter cities could create more absentee owners. Such cities are run by investors who live far away and seek to manage cities mainly for rents. These effects would be similar to the ramifications of developing the proposals of the Western Left Consensus.

Clearly, then, both the Conventional Wisdom and the Western Left Consensus are quite similar in effect at least. Future research could examine these issues more carefully. Indeed, the recent call for research on intergenerational inequality, focusing especially on upper-lower and lower-middle-income groups because this line of analysis is neglected in contemporary urban studies (MacLeavy & Manley, 2018) could usefully include the question of land and its transformation in our technologically mediated urban life. Likewise, policies on technology, industrialization, and employment could benefit from considering new ways of urban land reform that can break the cycles (see Figure 5.1). Overlooked by advocates of the Conventional Wisdom, including advocates of the SDGs and those who contend that technology is the definite solution to the multiple crises of the commons (e.g., O'Brien, 2018), and the Western Left Consensus advocated by modern political economists, these dynamics must lead to a stronger focus on land.

This focus is all the more important when we consider that most of the technological boom, in fact, depends on the extraction of rare minerals, which are used to manufacture key parts of technology (Klossek et al., 2016). Europium, for example, is used in the production of computer monitors, and neodymium is used in the production of smart phones and hard drives (Mertzman, 2018). The platinum mines in South Africa generate the ingredients for the production of catalytic devices used in cars to reduce emissions, so platinum, technology, and global political economy are interdependent (Yang, 2009). On the other hand, the extraction of minerals is strongly dependent on available technologies, such as advances in separation technology (Ge & Lei, 2018). Indeed, more oil is now being drilled as previous technological barriers are rapidly overcome (Frankel, 2007).

But therein lies another major challenge: would the oil fields be over-exploited? Could this "tragedy of the commons" be averted by unitization and governance by markets controlled by TNCs as claimed by leading economists such as Oliver Williamson (1981, 2002, 2009)? Or, as we face a common global environmental challenge, fuelled by the burning of oil, should we refrain from using oil and instead leave it in the ground for a cleaner common planet?

Chapter Six

Oil

Introduction

Drilling oil has been central to political economy and geopolitics. Along with fuelling global consumerism, the oil industry is also the focus of the global struggle for a clean and green planet. Most recent concerns about oil have focused on fracking in the Global North, especially in the United States; China's soaring consumption, which has made it the second-largest oil consumer in the world; the crisis in the Middle East, centred on how oil price volatility is going to shape the region's political economy ("To the Last Drop," 2019); and the socioecological crisis in the Amazonia petroleum fields (Cepek, 2018).

These concerns underpin the debate between the Conventional Wisdom and the Western Left Consensus. Whether to keep drilling oil commercially or to leave it in the ground is the historic dispute. Both sides accept that there is "a tragedy of the commons," but they propose different measures to combat it. On the one hand, for the Conventional Wisdom, *governance* by oil TNCs can address this tragedy. The Western Left Consensus, on the other hand, contends that drilling oil is inherently a resource curse (Goodman & Worth, 2008). So oil drilling, indeed, all drilling of fossil fuels, must end and be replaced with developing renewables.

This chapter questions the claims of both sides of the debate. Not only are the fundamental assumptions about a tragedy of the commons flawed, but the analytical approaches of both paradigms are questionable. They are neither historical, dialectical, nor evolutionary. Neglecting the institutional foundations of the economy, their explanations confuse symptoms for causes. I develop a more radical third position that takes into account multi-scalar property rights and, hence, provide new explanations and possible resolutions of the current socioecological crises. The rest of the chapter is structured into five sections examining

"The Oil Fields," "Coal," "Renewables," "Nuclear," and "Energy Sovereignty." I draw on examples primarily from Africa, partly because these examples are neglected in the commons debates, partly because these examples require further analysis (Renom et al., 2020), and principally because they substantially enable us to consider but ultimately transcend the Conventional Wisdom and the Western Left Consensus and instead to advance a Radical Alternative.

The Oil Fields

In 2012, Africa oil pools accounted for 5 per cent of global oil reserves and 7 per cent of global oil production. One in three oil discoveries in the world is in West Africa. This oil is mostly high grade and attracts considerable investment interest. Around five hundred companies are prospecting or drilling for oil in West Africa, including major Australian oil companies (Obeng-Odoom, 2015b; West Africa Oil Watch, 2014). The key oil producers in West Africa are Nigeria and Equatorial Guinea, recently joined by Côte d'Ivoire, Ghana, and Sierra Leone.

The experiences of oil cities in Africa are often overlooked in the literature on oil in Africa. However, it is important in analysing oil to pay attention to cities because focusing only on the nation-state – as has been the practice in research on oil – can be misleading. Some urban economies are driven by the oil industry, although the nation-states within which they are embedded are not oil nations at all (Barrionuevo & Peters, 2019). Oil cities are crucial sites of analysis because they provide the space for both oil extraction and consumption, including burning oil to power cars and to fuel urban industrial activities. Generally, "extraction accounts for only a small fraction of emissions associated with each barrel of oil; 70–80% occur when the customer burns it" often in cities, so the rise of oil cities must represent more than a simple nod ("Special Report," 2019, p. 9). Abidjan and Sekondi-Takoradi in Côte d'Ivoire and Ghana, respectively, require more careful analysis. National capitals, even where they are not the site of oil extraction, also benefit as the seat of power and, hence, must also be of serious interest. That is evidently the case of Freetown in Sierra Leone, where the power over exploration and even self-determination, perhaps excessive amounts of it, has been concentrated in the presidency (see, for example, Obeng-Odoom, 2014b, 2020).

Oil reserves cannot be assumed apriori to generate prosperity. Indeed, the debate, often cast in terms of resource curse and blessing, is, properly construed, disagreements about whether the oil fields are also sites of a tragedy of the commons. *Oil fields,* in this context, ought to be understood in two respects. The first is neocolonial; the second, imperial. Historically, Africa has been socioecologically constructed as a field of energy: slaves, coal, and oil (Ross, 2017, pp. 202–205; Showers, 2014). In this sense, colonialism was, in essence, an "enlightened" attempt to address a tragedy of the commons, an attempt to shed light on a savage continent where resources were *terra nullius*. Colonists such as Britain and France, therefore, aided their private interests – much like what happened in governing Africa's water resources (as discussed in chapter 7) – to govern the commons.

The private transnational corporation, then, was neocolonial from the beginning. It aided and abetted this massive socioecological experiment on colonization. Imperial tragedy of the commons is pseudo-scientific. That is, when oil fields are "discovered," forced "scientific" questions are raised by advocates of the Conventional Wisdom. Such questions include whether overproduction might ensue and concerns about the most efficient ways to extract oil from its field (Balthrop, 2012; Kim & Mahoney, 2002; Libecap & Smith, 2001; Libecap & Wiggins, 1984; "To the Last Drop," 2019). The key issues are these: "The resource is not exploited efficiently in a spatial sense because correlative rights are infringed upon …; The resource is not exploited efficiently dynamically because extraction occurs too early relative to the price rule; … There is physical inefficiency because rapid extraction damages the reservoir. There is economic inefficiency because too much of the rents from the resource are dissipated in the variable factors of recovery" (Balthrop, 2012, pp. 1–2). As I have shown elsewhere (Obeng-Odoom, 2019c), these two notions of *oil fields* are interlinked. Both enure to the benefit of imperial states, both enjoy the benefit of the use of the imperial military institution, and both lead to a further strengthening of an oil-dependent military to "discover," control, and maintain more oil fields.

The Conventional Wisdom is centred on the idea that corporate-led oil drilling – whether under colonialism or imperialism – is needed for development. The specific strategy, from this perspective, is to rely on TNCs. These governance units, as TNCs are regarded (see, for example, Williamson, 1981, 2002, 2009; "What Companies Are For," 2019), positively influence oil-based development through corporate social responsibility, collaborative planning with local governments, and wider programs of social licensing, along with being accountable and dynamic. (For a commentary on these specific mediums, see Marais et al., 2018; Obeng-Odoom, 2018; "What Companies Are For," 2019). This "collective capitalism" ("What Companies Are For," 2019, p. 9), according to the Conventional Wisdom, addresses the "tragedy of the commons." Advocated by organizations such as the African Union and the African Development Bank, Development Centre of the Organisation for Economic Co-operation and Development, United Nations Development Programme, and United Nations Economic Commission for Africa (UN ECA, 2011), this view holds that

> Effectively harnessed and well managed, Africa's resource wealth could lift millions of people out of poverty over the next decade. It could build the health, education and social protection systems that empower people to change their lives and reduce vulnerability. It could generate jobs for Africa's youth and markets for smallholder farmers. And it could put the region on a pathway towards dynamic and inclusive growth. (African Progress Panel, 2012, p. 11)

Contrary to popular perceptions that regard this standpoint as naive, it has strong economic foundations. For example, the staples thesis, developed by Canadian radical political economists, posits a strong theoretical connection between resource abundance and development. The successful Canadian economy is based on mineral extraction (Mills & Sweeney, 2013; "To the Last Drop," 2019). Recent advances in geography (Arias et al., 2014; Breul, 2019), anthropology (Richardson & Weszkalnys, 2014; Weszkalnys, 2018), sociology (Davidson & Dunlap, 2012), and applied local economics (Marais et al., 2018; Robbins, 2012) all support the view that Africa can use its oil commons for prosperity without disease, poverty, and ecological pillage. Geographers use the notion of clustering to analyse how geography explains the development of oil industries and how the development of those industries, in turn, shape the geography of oil industries.

They seek answers to questions such as in what ways markets and governments at different scales (such as federal, provincial, and municipal) collaborate or compete in stimulating the development of localization and urbanization economies, and how such interrelationships evolve over time. Such analyses are exemplified in T. W. Cobban's book *Cities of Oil* (2013), which shows how Canadian public policy used by the oil industry transformed and improved social conditions in southern Ontario specifically and Canada generally. Anthropologists have concretely documented many cases of mining transforming settlements positively, as have modern sociologists who take a more critical view of predictions of social disruption made by their forebears; specialists in local economies stress positive forward and backward linkages between the extractive industries and local economies (see, for example Lawrie et al., 2011; Malik, 2019).

Yet attempts to adapt this approach by African states have been widely discredited. Critics list a litany of examples of how oil-rich countries have failed in providing social services and economic prosperity, and how oil resources make regimes despotic. These criticisms feed into and are shaped by two conceptual approaches: "resource curse" and "rentier state" (Collier, 2009). In essence, these characterizations are reflections of the commons arguments, which essentially posit a deterministic view of how corruption and individual greed are derived from the governance of oil resources in Africa. Indeed, in 2013, Isabel dos Santos, daughter of the leader of the Angolan petrol state, became the first African woman to be listed in the Forbes List of the *nouveau riche* in the world. By 2020, she was the focus of global fraud investigations. Legal proceedings on corruption have been initiated against Teodoro Obiang Mangue, son of President Obiang Mbasogo of Equatorial Guinea (Africa Progress Panel, 2013). Many African governments, politicians, and their cronies are corrupt, of course – much like many presidents and politicians in powerful countries in the Global North. Individuals on the continent – as elsewhere – have become fabulously rich within a short time. The wealth of Nigeria's oil barons can be disturbing in its size and distribution over time. The situation is similar elsewhere in Africa and the rest of the world (see, for a general discussion, Wiegratz, 2019).

However, these social problems do not imply that Nigeria is a lost cause. They do not even show that Africans are possessed by a "culture of corruption." They cannot, therefore, imply the death of the African state, whose place in economic management must be taken by capitalist markets. The stimulating work of Jìmí O. Adésínà (2012) in *Mineral Rents and the Financing of Social Policy* (Hujo, 2012) is revealing. This analysis shows the situation before and after oil discovery in Nigeria, pre-structural adjustment mineral policy, and the role of the IMF and World Bank policies in weakening the Nigerian state post-adoption of the structural adjustment programs. Adésínà demonstrates that the specific role of oil in explaining the Nigerian condition is exaggerated. In fact, until the weakening of the Nigerian state through the imposition of Bretton Woods policies, mineral revenues were beginning to support social protection and economic programs for structural change. That developmentalist orientation itself was the product of years of failure in attempting to court private sector interest in supporting social programs and wider economic changes. As Adésínà demonstrates, the troubles in Nigeria are as much – if not more – the outcome of years of travelling along the neoliberal path dictated by Nigeria's "development partners."

In *Africa: Why Economists Get It Wrong* (2015), Morten Jerven explains why this important insight – and many others – elude most economists writing about Africa. First, much economics research is ahistorical, so it lacks context and nuanced perspective. Second, and partly deriving from the first, the misinformation from the lack of insight into the African context is compounded by a methodological approach that places the utmost faith in numeric data whose very reliability is questionable. Third, simplistic correlations and dubious causation analyses without detailed, long-term study of growth and change characterize most economic research on Africa. Such studies produce partial and outright wrong conclusions.

Take corruption. This social canker is typically said to be the bane of the African state when, in fact, it is even more strongly rooted in the market (Wiegratz, 2019). Brutal ambition typically drives capitalists to compromise the state, and some elements in the state view capitalist processes as avenues for personal enrichment. There is a strong connection between markets and corruption, demonstrating the moral fragilities of the market. Thus, in analysing the pervasiveness of corruption in Ugandan society, Jörg Wiegratz (2016, 2019) discounts the Conventional Wisdom about African culture as the breeding ground of corruption. Instead, he points to the economic system itself and its supporting institutions when he notes that:

> Economic trickery, fraud and crime are widespread in today's global economy. Affected are not only the markets for arms, drugs and human beings but also everyday economies for legal goods and services in sectors such as agro-business, manufacturing, food, health, banking, accounting, emissions trading and others. Fraud has invaded almost all societal subsystems and has become a major issue in both the Global North and Global South ... Fraud is not just significant in the poverty-stricken parts of the global economy, i.e. among "the poor," but also with its wealthy, powerful and hyper-rational centre: the boardrooms and offices of firms that dominate both Wall Street and high street. (Wiegratz, 2016, p. 1)

Yet the existing emphasis on "good governance," stressing "transparency" and private sector management as the best way to manage Africa's resources, invariably leads to the expansion of markets as the state is subjected to a discourse of corruption (Obeng-Odoom, 2010b). Attention is diverted from big corporate corruption to improprieties among state officials, although the two are intimately related. But even more fundamentally, the current analysis disregards a central pillar of progress and poverty: economic rent and the use of oil for energy sovereignty.

Key questions include whether governments view resources as common property, whether citizens are getting a fair share of oil revenues, and how to determine the best way to share oil revenues (Churches, 2009; see also Obeng-Odoom, 2015e, 2020). These questions are often overlooked in mainstream discussion about oil in Africa. Yet they need to be urgently asked to replace simplistic questions that lead to dichotomous analyses framed around euphoria (resource blessing) and pessimism (resource curse). Should rent created in and largely by African societies be captured by foreigners and exported to foreign lands? "Should Africa, again, be expected to carry Europe's burdens? and, Should the world's most energy-deficient societies be coerced into providing energy to some of the most energy profligate and well-endowed societies?" (Showers, 2014, p. 311).

Where TNCs have monopolized resources, the evidence has pointed to economic, social, and ecological challenges. Following Obeng-Odoom (2018), questions of urban poverty and, even more fundamentally, rising urban inequality can be raised. Problems of exploitation of labour – as with all other firms – with oil TNCs is well known, but, in addition, the labour aristocracy in such TNCs must be emphasized. Its heavier reliance on imported labour that earns much higher rewards than locals with similar experience makes it more culpable of labour aristocracy. So, is the evidence about people who have lost their housing and land or have been forced to relocate to slums because of the spatial needs of transnational companies? What is less emphasized is that TNCs have now become planning authorities in many such cities. In the oil cities of Sekondi-Takoradi in Ghana, TNCs fund the urban plans, which, in turn, tend to favour them. Giving planning permits for urban development that favours TNCs is common. Indeed, although it is now well known that atmospheric pollution in cities in the Global South is on the rise, it is not well emphasized that the reason is not just because TNCs supply more cars but also because the fuel sold to the South is more toxic.

These findings (Obeng-Odoom, 2018, 2020) also show that markets have been important in the rise of TNCs in cities in the Global South, but so have other institutions including the state and the army, and economic growth is expanding but it is not generating the "convergence" and "development" predicted by the stages of growth spatial economics treatise (Obeng-Odoom, 2016b). Cities in the South are becoming more unequal in the process of resource-based (urban) development. The standard explanations, including resource curse claims, are distractions, but do advocates of the Western Left Consensus provide a more convincing explanation?

The Western Left Consensus typically argues that oil and, with it, coal – indeed, all fossil fuels – must be left in the ground as drilling inevitably fuels capitalism's onward march to environmental destruction. Renewables are the preferred alternative, although nuclear power is increasingly becoming proposed as a middle alternative, too. Globally, the most well-known example of the victory of the Western Left Consensus was the Nasuni-ITT Initiative in which the Ecuadorian state was convinced to leave precious reserves of oil in the ground in exchange for compensation. Although this initiative was eventually abandoned, the calls for Africa to leave oil in the ground have become relentless. According to the Nigerian ecological activist Nnimmo Bassey (2012):

> Africans need soil, not oil. The environment is the cradle in which Africans are nurtured. Crude oil extraction has effectively uprooted the people from the soil. It has polluted their waters and poisoned their air. (p. 121)

Friends of the Earth Nigeria, which Bassey leads, recognizes that stopping oil production would throw oil-dependent countries into a crisis and, hence, offers a carefully developed alternative, based on a three-step logic. First, calculate how much oil revenue can be obtained per capita. Next, ask the citizens to pay this amount in taxes so that the state will get the same amount of revenue. Then, as not everyone can pay these taxes, share the remaining "unpaid taxes" among those who can shoulder more. Rich civil society groups can also support the initiative by buying oil under the soil – without actually receiving it, a quasi-Ecuadorian strategy. Leaving oil under the soil has many benefits, according to Bassey's group: it is a sure bet against flaring, pollution, and oil-extraction-related climate change. It will put an end to the displacement of local communities, nip corruption in the bud, put an end to violent conflicts, and maintain a clean environment (Bassey, 2012, pp. 127–129).

The analysis of the Western Left Consensus is problematic on several grounds. Its attempt to separate the political economy of oil from society more widely erroneously frames the oil fields in ecological terms, with social implications. In fact, the oil fields have always been socioecological in form and substance (Obeng-Odoom & Bromley, 2020). Separatist analysis of oil extraction and processing from its preferred alternatives of renewables is another analytical faux pas. As noted in chapter 5 on technology, the oil industry is inextricably linked to renewables. Indeed, there is a long and established tradition on the impossibility of a technological solution to the mining problem, developed originally by William Stanley Jevons (1906/2012) in *The Coal Question*. In this historic book, Jevons showed the dependence of renewables on fossil fuel. This analytical problem aside, there are also ethical and political-economic problems with the Western Left Consensus. As leading African scholar Julian Agyeman has consistently argued (see, for example, Agyeman, 2008, 2013), the most vibrant and most visible advocates of this Western Left Consensus are various ecologists and activists in the West, who organize around and often tend to deify the idea of "sustainability" as solely ecological.

From this perspective, sustainability is a singular vision, a non-material separatist ideal, one developed, maintained, and advocated by Western ecological activists. Marcia Langton, the Indigenous anthropologist and a leading Indigenous scholar (Langton, 2010), and Wangari Maathai (2004, 2011), African ecologist and Nobelist, have shown that much of this Western Left Consensus neither reflects black, Indigenous needs nor seeks to develop black Indigenous economies and potential. The Western Left Consensus is weak on the Indigenous idea of "Just Sustainabilities" (Agyeman, 2008, 2013; Nagendra, 2019). Indeed, writers who have typically appealed to just sustainabilities have been Southern writers (e.g., Ahmed & Meenar, 2018), with a few respectable exceptions such as Herman Daly, John Cobb Jr., Clifford Cobb (e.g., Daly et al., 1994), Erik Swyngedouw (e.g., Swyngedouw, 2015), and Frank Stilwell (e.g., Stilwell, 2017) who have been critical of the top-down colonizing claims of Western Left Consensus. Others have been insensitive to global inequalities.

Therefore, leaving resources in the ground, the organizing logic of the Western Left Consensus, is a product of misdiagnosis, a hangover of left-wing analysts' long-standing unwillingness or inability to fully engage the specific struggles of Africans in the global system. That paradigm reflects the left's *continuing* insidious belief that Africans need "enlightenment" patterned after left-wing thought (Diop, 1967; Fanon, 1961). Conceptually, this Western Left Consensus is based on problematic frames of thought. As shown by K. W. Kapp (1971) in his classic book *The Social Costs of Private Enterprise,* socioecological challenges arise from specific social relations developed, maintained, and expanded in capitalist societies (see also chapter 4). A decolonized explanation can be found in the private appropriation of socially created rent through the use of covert and overt force, whether under capitalism or any other system. To focus, then, on keeping the oil in the ground raises serious conceptual problems. Coupled with its rather insidious colonial "ecological enlightenment" (Adams & Mulligan, 2003, pp. 1–15), it can be safely replaced with a Radical Alternative based on developing resources as a common. Writing about African processes and the attempt to discredit them by the appeal to subtle colonial enlightenment practices, Cheikh Anta Diop (1967, p. xiv) notes that "all the head-long flights of certain infantile leftists who try to bypass this effort can be explained by intellectual inertia, inhibition, or incompetence." The African alternative, then, is to be defended.

In this Radical Alternative, the recognition and concrete pursuit of the wider African notion of "just sustainabilities" are crucial, as is the awareness that the pursuit of just sustainabilites requires the embrace of what Africans call "just transition," an idea that recognizes the peculiarities and complexities of African societies and the contradictions of moving away from oil-based developmentalist models (Marais et al., 2018). They stand in contrast to the narrow separatist notions of "just transition" and "just sustainability" used in European environmental discourses (see, for a review, Heffron & McCauley, 2018).

The rest of the chapter seeks to develop this Radical Alternative by first systematically considering the Conventional Wisdom, and then examining the Western Left Consensus. I use this narrative arc for coal, renewables, and nuclear power, all of which leading analysts such as Amartya Sen (2015) consider as "newer challenges" and "priorities of development research" (pp. 8, 15).

Coal

Coal continues to polarize opinions. Unlike oil, the coal debate can be particularly hot (James, 2019). On the one hand, its proponents see growing opportunity for economic transformation. The Conventional Wisdom is not oblivious to the environmental problems of coal mining. Rather, recognizing these difficulties, advocates make the case for "clean coal." Critics see a clean environment. However, only "zero coal" can guarantee such a world (James, 2019). According to the Africa Progress Panel (2015), "Coal is the dominant primary energy resource for the region, accounting for 45 per cent of total electricity supply" (p. 76). "Sub-Saharan Africa," the Panel continues," has abundant reserves of coal and oil. At current production levels, coal reserves are sufficient to meet demand for around 141 years" (p. 78).

The distribution of coal reserves and how those reserves have been used are, however, uneven. Mozambique has the potential to emerge as a major producer, with estimated reserves of 25 billion tons. Yet South Africa's share is far more substantial than the reserves of all other African countries put together. Indeed, alone, South Africa contributes about 6 per cent of the total stock of coal in the world and ranks sixth among the world's leading coal producers. It is coal that has propelled South Africa out of what Ayodeji Olukoju (2004, 56) calls the "epileptic power supply" problem in Africa. With coal, the recurrent power fluctuations common in many parts of the continent (Motengwe & Alagidede, 2017;

James, 2019), especially in Nigeria where children sing "up NEPA" when power is restored because they "never expect power always (NEPA)" (Olukoju, 2004, p. 56), is less known in South Africa. Together with coal-based prosperity, the South African experience can look promising for other countries on the continent, even if the issue of emissions remains an albatross around the neck of South Africa (Motengwe & Alagidede, 2017; James, 2019).

Across Africa, coal-fired power plants are rapidly being developed. Malawi, Zambia, Zimbabwe, Nigeria, and Senegal are, respectively, developing 300 MW, 300–600 MW, 600 MW, 1200 MW, and 1050 MW coal-fired power plants (Jacob, 2017). Tanzania, too, is planning to establish over two thousand major coal-fired power plants (Jacob, 2017). Asian demand for African coal is particularly high, sustained, and rising (James, 2019). Most of the coal mining projects in Africa are financed. As recently as June 2016, the chairperson of the Africa Union noted that coal will, if not should, form the backbone of Africa's energy renaissance (Jacob, 2017). There is, in short, a "surging appetite for coal energy in Sub-Saharan Africa" (Jacob, 2017, p. 343). Opportunity also beckons locally. Insufficient levels of electricity is such opportunity. Population growth may be a second. However, it is the drive for growth and the gulf between the wealthy Western nations and Africa that, for regional coal miners at least, secures the future of coal (James, 2019, p. 11).

Africa's coal question also magnifies the polarized debate on how to 'fix' coal in the world. On the one hand, resource optimists and growth advocates, such as Paul Collier of the Oxford Centre for the Study of African Economies and Donald Kaberuka, economist and former president of the African Development Bank (see, for a summary, Jacob, 2017; James, 2019), contend that more coal should be extracted from the ground, subject only to greater marketization. From this perspective, the emphasis is on better management of coal, that is, its price in the market, reduction of the emissions from its combustion, repair of the environmental degradation from extraction, and investment of returns such that economic growth that derives from its exploitation could justify its use as a source of energy. The net-zero emissions trading system ("Taxing Carbon," 2020) is a complementary strategy. Here, coal companies can continue polluting as usual, but they need to pay small farmers in the Global South to plant trees that will supposedly offset the pollution by these transnational corporations and companies. The sale of these carbon credits could be done either on a primary market or in the secondary carbon market, where those with extra credits can financialize and extract rent from land. This financialization of land is strongly aided by carbon engineering and the technological book discussed in the previous chapter. A triple win for markets, carbon, and technology, net-zero emissions trading schemes have received the imprimatur of *The Economist* newspaper, too (see, for example, "Taxing Carbon," 2020, pp. 56–59). The history of this approach to the Conventional Wisdom about the environment and how it became a central part of the Kyoto Protocol and the smorgasbord of global environmental instruments, including the Clean Development Mechanism and Joint Implementation can be found in recent scholarship (Bryant, 2019, pp. 73–97).

The emphasis on market forces and enclosure of the commons is anchored on the age-old idea of *the tragedy of the commons,* postulated by Garrett Hardin (1968). According to the tragedy of the commons idea, as explained throughout this book, when natural resources are held in common, they are overused and abused and they lead to pollution and exhaustion. State intervention is not ideal because the state is an inefficient manager and people prefer marketization. This view is made more explicit in the "open access exploitation thesis," but Hardin's ideas remain the silent underpinning framework for other theoretical perspectives too, including watered-down versions that recommend public-private partnerships. This emphasis largely explains the turn to governance, that is, considering the transnational corporation as a governance structure (Williamson, 1981, 2002, 2009) to solve resource problems, along with the rise of new institutional economics as the intellectual home of economic governance research (Boettke et al., 2012).

From this governance perspective, the market can be strengthened to enhance the near perpetual drilling of coal. According to the Hartwick rule, for instance, it is prudent for the resource-rich economy to invest the returns from natural resources in physical and human capital to lengthen the exhaustibility period and to prepare for the post-resource phase (Hartwick, 1977). Milton Friedman's permanent income hypothesis is the other idea that has triggered much interest in Sovereign Wealth Fund (Alagidede & Akpoza, 2015; Amoako-Tuffour, 2016). Similarly, Harold Hotelling (1931) offered the analytical foundations for the view that the price mechanism is the best way to ensure that exhaustible resources are exploited at a socially optimal rate. For Hotelling (1931, p. 173), the dictum "higher prices and lower rates of production" is a general condition or universal law, which acts as an invisible hand that guides mainstream economists today (for a critical discussion, see Rosewarne, 2011). To the extent, then, that ways can be found through the market for more coal to be mined "sustainably" (read as over a long period or by using less-polluting technologies) through investment in technology, human, and other physical capital, the mainstream economics view holds that the more coal there is, the better. Buoyed by donor insistence on the "Big Push" approach common in the work of mainstream economist Jeffrey Sachs (2010), the prevailing view is one of neoliberal policy forms such as Tanzania's "Big Results Now" developmentalist initiatives (Jacob, 2017). So much for the Conventional Wisdom and its ideas of sustainable coal or clean coal (MacDowell, 2017).

On the other hand, the Western Left Consensus advocates that coal should be left in the ground. Known as the zero-coal advocacy, this Western Left Consensus is part of a grand support for climate change policies. This position was delivered by President Kikwete of Tanzania during the nineteenth and twentieth sessions of the United Nations Framework Convention on Climate Change. In a broad array of programs, including the African Ministerial Conference on the Environment, there appears to be some official disdain for a red-hot African economy based on coal. This "official line" seems to support a green Africa without coal (Jacob, 2017). Nnimmo Bassey (2012), a well-known environmental activist, argues that the fossil fuel development model, in general, has failed to improve social conditions. Rather, this fossil-based development has been polluting and destructive of society and economy. Others such as E. G. Frankel (2007), writing more generally, contend that the world is moving away from fossil so it is better to move with the world. The implication is that coal has no future. African coal too will soon have no one to buy it. Either way, interest in the role of Africa in a post-fossil era is increasing, with publications such as *Energy Transition in Africa* (Simelane & Abdel-Rahman, 2011) and *Future Directions of Municipal Solid Waste Management in Africa* (Mohee & Simelane, 2015) leading the change.

According to the Western Left Consensus, renewable energy is the only compelling alternative. Yet this panacea can, in fact, be a poison. Advocates are typically inattentive to possible spatial mismatch between old jobs that are lost in the energy transition and new green jobs that will be created (Acey & Culhane, 2013; Pearce & Stilwell, 2008; Stilwell & Primrose, 2010). Indeed, the crucial question of whether technologies that are used to exploit renewable energies will allow successful transition that has less environmental impact compared to exploiting coal or leaving it in the ground still remains (Goodman & Rosewarne, 2015; James, 2019). Analytically, this third way overlooks the historical development of coal, a subject that has direct relevance to the viability of renewables because the coal question always included the renewables question, as an oft-forgotten major study, *The Coal Question*, written by English economist William Stanley Jevons (1906/2012), shows.

The Coal Question is a classic and major text for research on the political economy of natural resources. Many studies suggest that its insights (e.g., the geographical finitude of coal and the paradox of economizing on coal) are relevant today. E. G. Frankel (2007), for one, notes:

> The stage is now set for the eclipse of the fossil fuel age in global development. It will start with the rapid replacement of traditional coal and petroleum use in power plants and industry by cleaner burning gas and other fuels, as well as nuclear, wind, hydro, and solar power processes. We expect that within 20 years (2027) only about 10% of utility and industrial fuel will be petroleum and coal. In fact, it is expected that its use in power generation may be phased out completely before the middle of this century. (p. 4)

M. King Hubert (1974) is much better known and is more widely credited with the idea of peak fossil era. Notably, Jevons's work predated Hubert's. Yet a review of Jevons's work is needed because there is much confusion about its central contribution to the body of thought on fossil fuels and the exploitation of related resources. An early attempt to review Jevons's contribution was made by J. M. Keynes (1936), but it was light on coal, and even this small mention of the coal question contained a fundamental error. Keynes claimed that Jevons's *The Coal Question* (1906/2012) was merely about the exhaustibility of resources and that Jevons had not taken into account technology. Neither of these was correct, however; Jevons's argument was more sophisticated than a simple focus on the exhaustibility of resources, and his entire book was a challenge to technological fixes of the coal question, leading to what is now widely called the Jevons paradox. Yet, surprisingly, Jevons's thoughts are sometimes misrepresented even in relation to the paradox. For instance, Richard York (2006) argues that Jevons never took into account capitalist pressures; but as I will show, Jevons did consider these forces. Political economists have simply buried Jevons, accusing him of unleashing mainstream economics upon the world – without acknowledging Jevons's pioneering contribution to a limit to growth thesis and his advocacy for using the state as an institution for redistribution and ecological change, as he did in his classic work *The Coal Question* (1906/2012).

Our knowledge of *The Coal Question* (1906/2012) remains partial, sometimes even incorrect. Generally, "the emergence of competing narratives about energy ... calls for deeper inquiry into the political economy of energy transitions to establish a better understanding of the competing interests at play and the actors involved" (Jacob, 2017, p. 353).

It is this gap that the rest of the chapter seeks to address. In doing so, I argue that the analytical foundations of the debate about coal mining in Africa tend to be buried or sacrificed for the advocacy of certain policy choices centred on the prevailing binaries of the Conventional Wisdom and the Western Left Consensus. Attempts to bridge the polarized positions – often centred on the development of renewables – are contradictory. Dissatisfied with these three positions, the rest of this chapter revisits the coal question by seeking to develop W. S. Jevons's path-breaking treatise, *The Coal Question* (1906/2012), an analytical critique of, and alternative to, the existing state of knowledge. Jevons's work is insightful but incomplete because it is weak in its grasp of property relations in the coal industry, which, in the case of Africa, are highly monopolistic. A stronger framework must view as nested the analysis of the relationship between coal and technology, the implications of the historical development of energy for the current interest in renewables, and the relationship between ecological and economic questions. From this perspective, Africa can usefully seek self-reliance in the use of its coal resources. The continent can also change existing property relations, avoid the technological approach of ever-increasing extraction, and jettison the ideology of growth. Degrowth, however, is neither consistently demanded nor systematically theorized by the Western Left Consensus. As a broad-based movement (see Research & Degrowth, 2010; Borowy & Schmelzer, 2017), however, this inconsistency is understandable.

Renewables

Yet the contradictions in the Western Left Consensus ought to be discussed, if the interest in investing in renewables is to be taken seriously as one alternative. As Kate Showers (2014, 2019) has shown, *renewables* can be defined as anything that is not based on fossils. Renewables usually have a *green* side to them, but their greenness could range from biofuels to green cars. Technologies that rely on local inputs or that agronomically use less land are also considered as part of renewables. So are slaves, especially black slaves: they are both non-fossilized and green, as their use has apparently no impact on the environment. There is, the argument goes, abundant, cheap, and marginal land for growing renewables and for enabling people to work the land sustainably (see, for examples of such uses, Shrader-Frechette, 2011; for a wider discussion of such uses, see Exner et al., 2015). These renewables are, of course, also advocated by some proponents of Conventional Wisdom (e.g., Collier & Venables, 2012; Holstenkamp, 2019) when they metamorphose into advocates of green revolution.

That is hardly where the contradictions in the Western Left Consensus end. Many proponents of renewables implicitly advocate growth as usual, with the only difference being the absence of fossil fuels. In their place, new technologies that use renewables are advocated. Indeed, such renewables are praised for being cost-effective, compared with dirty and expensive fossil fuels and nuclear power. That is evidently the argument made by Kristin Shrader-Frechette, a strong advocate of renewable energy. In her book *What Will Work* (2011), she puts forward a three-part case for renewables. First, renewables are "cheap." Second, they are a viable enterprise for profit. Third, they can be used to power green cars and other low-carbon technologies (see Shrader-Frechette, 2011, pp. 188–211) in ways that big business prefers (see Shrader-Frechette, 2011, p. 31). In other words, renewable energy development is both community empowering and commercially viable. So, we are back to profit-led growth.

Yet while it has been argued that "renewable energy technology is not just another option for the continent but the only option" (Sanni et al., 2014, p. 253), this approach is hugely controversial and even unethical. Not only is it insensitive to the controversial historical memories of the enslavement of Africans to provide renewable energy to the Global North (Showers, 2014, 2019), but Africa's contribution to the global problem is also rather insubstantial (even if its growing and the overall effect on Africa of a crisis of climate will be devastating). So to call on it to make, in essence, similar adjustments as the world's powerful polluters is rather questionable – even more so when it is *the underlying property relations*, not only extraction per se, that generates such problems. Historically, similar lines of problematic reasoning were used to justify the enslavement of Africans considered naturally abundant and strong to provide renewable energy for the rest of the world (Showers, 2014, 2019).

In any case, the production of renewables has generated mass displacements through widespread land acquisitions for renewable projects. In addition, solar requires substantial energy and non-renewable resources to make the solar panels. Wind farms use enormous amounts of concrete for the bases (which requires mining), large amounts of steel for the towers, petroleum products for plastics, and significant amounts of fuel for constructing the huge wind farms. According to Michael Fox (2014), "Numbers vary widely for these activities, but a range in the amount of CO_2- equivalent greenhouse gases in grams produced for each kWh of energy is 9–21 g/kWh for nuclear, 10–48 for wind, and 100–280 for solar" (p. 105). Coal, in contrast, "produces 960–1,300 and natural gas produces 350–850 g/kWh" (Fox, 2014, p. 105). Displacements and environmental problems with renewables can be explained in other ways, too. Most wind and sun alternatives are currently provided by the TNCs. Their projects tend to be on land, which is presumably empty, unused, free, or marginal. In practice, this land is part of the commons. Parcelling it out for private TNCs is, therefore, a systematic transferring of common wealth into private hands. Sunlight and wind, key inputs in the renewable alternative, are also regarded as free. So TNCs extract rents from privatized land. As Global South renewable alternatives are mainly investments by wealthy foreigners and foreign-based businesses, investors become absentee owners (for further discussion of this Veblenian concept, see Obeng-Odoom, 2020). In the name of developing renewables, a rentier class develops. These companies claim interest in sustainability, but their primary responsibility is to the class of absentee owners. Satisfying these invisible landlords comes at major opportunity cost: using land that was used for food or can be used for common food as an investment vehicle for wealthy, foreign landlords.

The experience of Egypt's Benban Solar Park, a major global solar farm, is a case in point (EcoConServ Environmental Solutions, 2016). The farm is supported by the international development groups. With more than 40 fields, Benban Solar Park's many private investors include Rising Sun, with several foreign investors based, among others, in Dubai. This absentee ownership creates what *The Economist* calls "a dangerous gap: The markets v. the real economy" (2020, pp. 7, 25). Local landlords also benefit. The value of land near these renewable farms tend to rise too, putting more socially created rent in the hands of private landlords. At a much more fundamentally deep level, the case of the Western Left Consensus has serious analytical limitations.

Analytical Limitations of the Coal–Renewables Arguments

The analytical foundations of existing challenge to coal are contestable. Currently, the coal question tends to be complicated by a property rights system that makes it possible for the rich to hide behind the poor, weaker races, classes, and genders to claim that coal mining helps the poor. So, analytically, overlooking the nature of property rights in space and time inhibits systematic explanation (Wang, 2019). In practice, weaker races, classes, and genders in peri-urban settlements and rural areas are overworked in the coal industry and yet are poorly paid. These coal workers also live the inhumanity of coal waste when their settlements are targeted as dumping grounds for coal TNCs (Talukdar, 2017).

Such matters do not disappear simply because coal is left buried in the ground. Indeed, the renewables industry is becoming highly exploitative along property lines. Labour is both excluded and exploited, unvalued and undervalued on the many green farms in Africa. Common water is being slowly enclosed to produce renewables. Communities are being evicted, and within them, inferiorized races and genders are the worst affected (Akiwumi, 2017; Chiweshe, 2017; Elhardary & Obeng-Odoom, 2012; Kuusaana, 2017; Obeng-Odoom, 2013f, 2016e). In the name of renewables, in developing clean, green, and sustainable energy, Africa is slowly returning to the days of slavery where people were used as sustainable energy (Showers, 2014).

Without decolonizing these positions, the separatist treatment of development and nature in the existing Western Left Consensus, itself led by Europeans and Western activists (Obeng-Odoom, 2017b), will cause more harm than good. The demands are often at variance with what African movements are seeking (see, for example, Adam Branch and Zachariah Mampilly's *Africa Uprising: Popular Protest and Political Change,* 2015; Rodríguez-Labajos et al., 2019). Simelane and Abdel-Rahman (2011) strongly argue for alternatives that meet three criteria. First, the alternative must be widely accepted by local people. Second, the alternative must help to build local capacity, and, third, it must help to replenish the continent's natural resources.

The Western Left Consensus does not meet all these criteria. While this consensus focuses on the local, it neglects local needs that are at variance with its environmentalist creed. Even at the local scale, it is too mono-scalar and overlooks the importance of multi-scalar analysis and action. It is not clear, for example, what the state should be doing in a post-development, Western Left Consensus world. Questions about social protection of the aged, the weak, and the maimed – better handled at multiple scales – get insufficient attention in post-development mono-scalar world. The emphasis on the small scale makes the Western Left Consensus vulnerable to criticisms that largeness is the problem, not the nature of social institutions. In turn, the Western Left Consensus can easily provide the justification for considering cities, for example, as problems and too much migration – whatever that means – as a crisis. Indeed, the tendency to praise the smallness of life in Africa overlooks the great cities of Africa such as Great Zimbabwe (see Grant, 2015, for more examples) and Pharaonic Africa (Diop, 1977). It is the antithesis of the approach in critical African political economy in which Africa is placed at the centre rather than the margins of the world; Cline-Cole, 2020; Cooper, 2014.

The Western Left Consensus tends to ignore black political economy and postcolonial analysis. Indeed, this Western Left Consensus seems to have arisen against and outside the rich body of knowledge developed by Indigenous scholars and others. Holistic concerns with the global system are neglected. Questions such as race – central to the entire idea of global development – are very poorly considered, and the South tends to be objectified, indeed essentialized.

Developing the radical challenge can begin with the work of W. S. Jevons, especially *The Coal Question* (1906/2012). In this book, Jevons developed an approach to natural resource economic analysis centred on the intersectional relationships among economic principles, ethical judgment for society and environment. His approach was historical rather than "the fickleness of statistical numbers" (Jevons, 1906/2012, p. 6). The book is divided into 17 chapters, encased in commentaries, and a crisp conclusion. Chapter 1 sets out the coal question as (1) the durability of coal and whether an economy can keep expanding forever, and (2) the fallacious thinking about alternatives in the form of renewables.

Chapter 2 is a review of the state of research on the topic at the time Jevons wrote the book. Chapter 3 is a critical account of the geological studies and features of coal. Chapter 4 is on the cost of coal mining, while chapter 5 looks at the price of coal and how it is determined, not to give a theory of coal price but to offer a historical analysis of trends. Chapter 6 covers the history of inventions and how they relied on coal. Chapter 7 is a comprehensive challenge to the idea that advances in technology can address social problems, including the exhaustibility of coal, and here Jevons introduces the idea that the technologies intended to save coal paradoxically expedite its exhaustibility.

Nevertheless, the solution is not to turn to renewables. Chapter 8 of the book launches an attack on alternatives such as electricity, solar, water, and wind. Chapter 9 discusses the law of uniform geometrical increase or the natural law of social growth. Here, Jevons wants to explain the exponential growth in our use of coal by using the logic in Malthus: we are a reflection of our past. The point, however, is that there are social forces that sustain this continuing pattern. The cost of continuing to use coal in the same way as we did in the past is the high price we now pay for new discoveries (see Jevons, 1906/2012, p. 198). Chapter 10 is about flourishing population growth and emigration driven by coal-induced world development in an attempt to show that people are mainly living a better life but that life will likely vanish if we do not stop using coal as we do. Happiness will give way to misery. It is an unusual thesis because he uses population numbers and marriages to gauge happiness.

Jevons acknowledged the role of coal in the transformation of the economic structure in Britain, and in chapter 11, he looks at the changing nature of industry and how coal has fuelled rapid industrialization. Chapter 12 provides an analysis of the alarming rate at which coal is consumed. Chapter 13 considers whether importing coal from abroad would solve the coal question in Britain and concludes that it would not because (1) importing costs Britain money as it is more expensive, (2) importing makes Britain dependent, and (3) importing harms the competitiveness of other sectors of the British economy. Finally, the advantage of linkages is low if Britain imports.

Chapters 14 and 15 are comparative, with the former analysing the coal reserves of other countries and the latter examining the trade in iron. In chapter 16, Jevons argues that if Britain continued to expand, its expansion would undercut its global standing. Why? Over-selling leads those source countries to develop, which in turn attracts Brits to emigrate. Together with local population, those countries can develop the industries to compete with Britain. Chapter 16 is also a methodological chapter because it says that we cannot discuss Britain in a vacuum. We need to consider it in relation to other settlements. The chapter is nationalist, though, and even glorifies colonialism. Chapter 17 puts forward the case of the only reform that Jevons argued would work: the reduction or elimination of the national debt. Doing so would (1) increase production capacity, (2) save posterity from future difficulties, and (3) reduce the excessive economic growth that has harmful implications for the society, economy, and environment.

Three key ideas in Jevons's *The Coal Question* (1906/2012) require particular emphasis because they are relevant to the contemporary political-economic analysis of coal in Africa. The first is the so-called Jevons paradox: the idea that more technology aimed at economizing coal can save coal from exhaustion is, in fact, the reverse. More technology increases the likelihood that more coal will be used, and as coal is not renewable, technology can instead expedite the exhaustion of coal. Technology leads to the production of more with less; in the end, more coal will be used to produce more and more. Indeed, as the price of coal falls, products made from coal become cheap, more coal-powered products are demanded and produced, and more coal is needed. And profit-making capitalists seek to extract maximum coal for use and sale to enrich themselves.

It is important, then, to dwell a bit more on this idea. Jevons summarized the debate this way: "It is very commonly urged, that the failing supply of coal will be met by new modes of using it efficiently and economically. The amount of useful work got out of coal may be made to increase manifold, while the amount of coal consumed is stationary or diminishing" (1906/2012, p. 137). Jevons conceded that at the household level, these claims might apply, but as the household use of coal is insubstantial (p. 138), the coal used by manufacturers and others requires the most attention. For that, Jevons noted: "*It is wholly a confusion of ideas to suppose that the economical use of fuel is equivalent to a diminished consumption. The very contrary is the truth*" (p. 140, emphasis in original).

Jevons gave examples about the introduction of technology and how it was supposed to reduce labour but in the end actually increased the amount of labour employed. The economy of labour that comes with the introduction of new machinery throws labourers out of employment at first. But such is the increased demand for the cheapened products that eventually the sphere of employment is widened. Often the very labourers whose labour is saved find their more efficient labour more in demand than before (p. 140). The example Jevons used is that of seamstresses. According to him, "Seamstresses ... have perhaps in no case been injured, but have often gained wages before unthought of by the use of the sewing-machine, for which we are so much indebted to American inventors" (1906/2012, p. 140).

He continues, "Now the same principles apply, with even greater force and distinctness, to the use of such a general agent as coal. It is the very economy of its use which leads to its extensive consumption" (1906/2012, p. 140):

> The number of tons of coal used in any branch of industry is the product of the number of separate works and the average number of tons consumed in each. Now, if the quantity of coal used in a blast-furnace, for instance, should be diminished in comparison with the yield, the profits of the trade will increase, new capital will be attracted, the price of pig-iron will fall, but the demand for it increase; and eventually the greater number of furnaces will more than make up for the diminished consumption of each. And if such is not always the result within a single branch, it must be remembered that the progress of any branch of manufacture excites a new activity in most other branches, and leads indirectly, if not directly, to increased inroads upon our seams of coal. (Jevons, 1906/2012, pp. 141–142)

What Jevons is arguing is not abstract. He cites the example of Scotland, rich in fossil fuel, especially coal: "*The reduction of the consumption of coal, per ton of iron, to less than one-third of its former amount was followed, in Scotland, by a tenfold total consumption, between the years 1830 and 1863, not to speak of the indirect effect of cheap iron in accelerating other coal-consuming branches of industry*" (1906/2012, p. 154, italics in original). In other words, "no one must suppose that coal thus saved is spared – it is only saved from one use to be employed in others, and the profits gained soon lead to extended employment in many new forms" (p. 155). It is here that Jevons offers detailed analysis of why economy is not economizing.

The existing alternatives to coal are not good enough. Jevons's argument is that such alternative power relies on coal anyway. In any case, they were overtaken by coal, which proved more portable, more user-friendly, and more reliable. Jevons put forward a bold proposal for progress: slow growth, a reduction of national debt, good communities, healthy ways of life, and people living without debt.

To pay the debt, Jevons recommended (1) tax inheritance ("legacy and success duties," 1906/2012, p. 449) or use the existing succession duties (as it was in the case in Britain at the time of writing, p. 451); (2) do not spend more than is earned; that is, be self-sufficient; and (3) give the proceeds of the tax to a separate commission to pay off the national debt. Revenues from inheritance tax must be put to debt reduction or elimination; otherwise they are wasted (pp. 451–452). Finally, the conclusion in Jevons's book put forward the case for the limit to growth thesis: seek a "stationary state" (see, for example, p. xxxi) "to secure a safe smallness" (p. 456) or risk the danger the country will "contract to her former littleness" (p. 459). Jevons closed with his famous wise counsel to Britain:

> The alternatives before us are simple. Our empire and race already comprise one-fifth of the world's population; and by our plantation of new States, by our guardianship of the seas, by our penetrating commerce, by the example of our just laws and firm constitution, and above all by the dissemination of our new arts, we stimulate the progress of mankind in a degree not to be measured. If we lavishly and boldly push forward in the creation of our riches, both material and intellectual, it is hard to overestimate the pitch of beneficial influence to which we may attain in the present. *But the maintenance of such a position is physically impossible. We have to make the momentous choice between brief but true greatness and longer continued mediocrity.* (pp. 459–460, emphasis in original)

Jevons's analysis is weakened by not taking into account social relations, especially property relations and global inequalities. He did not consider fundamental questions about coal contracts, the role of TNCs, and the private appropriation of socially created rent. What exactly are the social relations that characterize the coal industry? In particular, what property relations have been set in motion by coal at the national, subnational, and community levels? What land rights are being created/reshaped or destroyed? Is coal as a commons becoming more public, private, or in fact more common? What do these political-economic changes say about rent, and what, if anything, is the state doing to capture or have a fair share of rent for the public?

These questions are important if we are to analytically look at and actually take slow growth seriously. When coal is the private property of TNCs, slow growth is a pipe dream. Where coal is part of a commons and ownership is vested in community, however, the tendency is not to overproduce but rather to produce for self-sufficiency and a limited amount of exchange. Otherwise, leaving the system of property relations unchanged and showing, as Jevons did, that a slowdown is needed will not bring about needed changes.

Thus, it is also necessary to jettison third-way solutions of alternatives based on only renewables because, in fact, the true coal question is about contesting the ideology of growth, as Jevons showed. A more defensible answer to the coal question is to focus political-economic analysis on the harrowing unequal property relations that characterize extractivism; challenge neocolonial environmentalism, whose activities deflect attention from unequal social relations and naturalize social problems; and pursue self-determination based on principles of commons and commoning.

From the Coal Question to the Coal Commons

Creating new social relations is fundamental. Analytically, doing so will mean breaking away from both a central problem of development (patronizing the South) and a key setback of post-development Western Left Consensus (the objectification of the South). It is possible to remain critical of development by forging stronger analytical bridges among various intellectual traditions, such as post colonialism, political economy of development, land economics, stratification economics, and institutional and Georgist economics (Obeng-Odoom, 2015c, 2020; van Griethuysen, 2012), while seeking real-world social and ecological change through factual and analytically informed activism. Specifically, a march towards the commons can bring about these changes. One way to bring about the commons is to regard land and all fossil as commons to be used in the satisfaction of the common rather than for profit. Currently, only 101 companies, mostly of British descent and listed on the London Stock Exchange, dominate the coal industry throughout Africa. These companies control 3.6 billion tons of coal valued at $216 billion (Curtis, 2016).

Those in possession of coal can be asked to give it up, not through violence but by putting a tax on coal *extraction*. Such a tax recognizes

that the value of coal, of land, is created by labour locally and the specific value of coal is established *relatively* through pricing established
by labour production globally, through public investment, and through
speculation, not through anything intrinsic to coal. In that sense, the
value of coal must be commoned, for example, through the capture of
rent or the application of tax (see Henry George's *Perplexed Philosopher*,
1892/1981, Chapter 5). This does not increase the amount of tax in the
system because it also means untaxing labour.

Such a tax is just for three reasons. First, it gives to the public and to
labour what they created and hence overcomes the problem of a few
TNCs monopolizing value that they did not fully create. Second, this
tax relieves labour of exactions that are unjust and hence enables the
full enjoyment of the fruits of labour. Third, it overcomes a major structural mechanism for creating and sustaining inequality: private property and the creation and private appropriation of rent. It is this same
mechanism – private property in the commons – that drives pollution,
speculation, and sprawl, so the Georgist attempt to transform property
relations is well considered.

The government under such a taxation regime would become much
stronger and could use the rents from the resources to invest in public
goods and public works. As we learn in "The Functions of Government" in *Social Problems* by Henry George (1883/1966, Chapter 17), rent
as a public resource can strengthen the investment capacity of local and
national authorities. Empirically, international experiences analysed by
Brueckner and colleagues (2014) of commoning mining rents to massive afforestation programs and the nourishment of nature suggest that
the proposals by Henry George are plausible. If embraced more fully to
include a transformation in existing property relations, such proposals-
can bring about propitious socioecological change. Indeed, rents may
support sustainability directly or may be passed on to urban authorities
to determine how best to invest them in social services, public parks,
gardens, and public libraries.

These public spaces can become common spaces by switching from
top-down management typified by joint rights to bottom-up management characterized by equal rights in land. In this sense, governance
becomes an act of commoning rather than a system of majorities.
Everyone becomes part of managing the coal commons. The false antimonies between commons and the state collapse because the state is
based on the idea of the commons: a commons-based state or a state
that, in the words of Henry George (1883/1966), is a commons whose
"organization," "methods," and "functions" are "restricted to those
necessary to the common welfare, and in all its parts it should be kept
as close to the people and as directly within their control as may be"

(p. 171). In this system of the commons, coal for energy will be scaled down and eventually phased out. Removed from the hands of TNCs and put in the hands of commoners or their trustees, with whom they have close relationships, a system of 'common coal' can be established (compare to state systems based on joint rights in which political representatives are neither known meaningfully by citizens nor know citizens meaningfully; see George, 1883/1966, pp. 174–175). Such measures will enable more transparency and democracy while better reflecting the wants of the commoners. Simultaneously, such commons management can strongly contribute to creating what Jevons (1906/2012) called a "stationary condition" of low growth: "Our motion must be reduced to rest, and it is on this change my attention is directed. How long we may exist in a stationary condition? I, for one, should never attempt to conjecture" (p. xxxi).

What is crucial is that the state – at various levels organized as commons – plays a clear and decisive role in addressing the coal question. Reducing the national debt, as Jevons argued, is clearly important, but it falls under the functions of government; it is one of many steps that the government can take. Jevons was correct to look at the actual extraction of coal as the problem, not the management of coal or the attempt to influence the individual behaviour of people, not trying to do business as usual by appealing to sophisticated technology. Yet the problem is no better addressed by ignoring it or leaving coal buried. Rather, as examples around the world, including Alaska, show (see Widerquist & Howard, 2012a, 2012b), by boldly confronting the interlacing issues of unequal property regimes and analysing who currently owns what and where, who should possess new finds, and how resulting rent is accumulating and is likely to be shared will better help us to appreciate the intersections of ecology, economy, and society. Indeed, with the strong linkages between coal and growth (Motengwe & Alagidede, 2017), the proposals discussed are likely to change the nature of growth on the continent. Also, because "on a per unit basis, coal generates roughly twice as much CO_2 as natural gas" (Africa Progress Panel, 2015, p. 86), a reduction in coal extraction will also reduce existing amounts of carbon emissions. Leaving behind the false antimonies of blessings or curses is crucial to see the grave inequalities that characterize the extractive industries globally (Peters, 2017). These analytical insights could also be applied to nuclear energy.

Nuclear

Nuclear energy is also advocated in the Conventional Wisdom. In his widely acclaimed analysis of four hundred years of world energy, Richard Rhodes (2018) makes the case for nuclear energy. Not only is he supportive of nuclear energy because it decarbonizes energy, but also because nuclear is "radically decarbonizing" (p.332). For him, therefore, limiting global warming means decarbonizing with nuclear energy. For Rhodes, nuclear energy's record is analogous to wind and solar, whose reliability is not comparable to nuclear. In terms of public health, too, Rhodes notes, "Nuclear power's public health record more tan compensates for its few occupational accidents" (p. 336). Its limited air pollution combined with its extremely low greenhouse gas emissions and its 24/7 availability more than 90 per cent of the time make it easily the most promising single energy source available to cope with twenty-first-century energy challenges. Perhaps the best-known proponent of this Conventional Wisdom in Africa is Dr. Mojalefa Murphy. His case for a nuclear alternative for Africa rests on three grounds (for a detailed discussion, see Murphy, 2011). First, Africa has nuclear potential not only in terms of vast reserves of uranium but also in the scientific capacity to develop it. Africa's share of the world's uranium supply is 22 per cent, but it produces merely 0.5 per cent of the world's nuclear power in one power station in South Africa (Murphy, 2011, pp. 31–32). Second, the well-known potential problems of the use of nuclear technology to develop weapons of mass destruction, the secrecy that shrouds the nuclear industry, the risks of storage and disposal, and the devastating disasters caused by nuclear power cannot explain the underdevelopment of Africa's nuclear potential. Some of the most vocal opponents in the West who cite these problems actively mine uranium in Africa for their own strategic and national interests: "Africa … is experiencing an unprecedented influx of prospecting and exploring fortune hunters from Australia, Canada, Europe (including France and the United Kingdom) and other nuclear power countries such as China, Russia, South Korea and the United States" (Murphy, 2011, p. 31).

History (for a detailed discussion, see Murphy, 2011) clearly shows that it is mainly sabotage that has stifled the growth of this industry, and this neocolonialism has cost Africa significantly. One effect is the development of debt related to earlier investments that were undermined. A second effect is the underdevelopment of scientific capacity and training. A third effect is that, as with other resource problems in Africa, the continent has merely exported raw uranium for the benefit of others, without turning it into a resource for national and African-wide social transformation. For these reasons, nuclear, too, has become an albatross. If so, Murphy's third and final ground is that there is a clear case for Africa to use its nuclear potential to resolve its many challenges:

> Despite the shortcomings of the nuclear energy technology as well as safety
> and regulatory practices especially by the Fukushima disaster, nuclear

energy technology continues to hold a great potential for the reduction of GHG emission through other climate friendly applications other than electricity generation. The production of hydrogen which is widely used in agriculture, food industry, petrochemical industry and clean transportation among other applications, causes huge amounts of GHG emission, notably carbon dioxide due to the intensive use of fossil fuel to produce process heat for the high temperatures required in water electrolysis. The required high temperatures may not be produced by renewable energy systems as yet. When a nuclear reactor is used to provide process heat, the GHG emissions are completely eliminated … Thus nuclear energy will continue to play an important role in the 21st century green economy, in which Africa's competitiveness would not be enough if her nuclear technological potential would not be exploited. (Murphy, 2011, pp. 48–49)

Indeed, refusing to do so is immoral. South Africa and, by similar reasoning, Nigeria, the two economic giants in Africa, carry what Murphy (2011) calls a "moral duty … to provide leadership in the quest for the realisation of the African nuclear power potential" (p. 51). This line of analysis is intriguing in its attempt to deconstruct the double standards of the world order and to put forward a strategy of resource sovereignty.

This pro-nuclear alternative is gaining widespread support. Michael Fox, emeritus professor in the Department of Environmental and Radiological Health Sciences at Colorado State University, has made similar arguments. In *Why We Need Nuclear Power: The Environmental Case* (2014), Fox points out that the concerns about excessive radiation imperilling the health of communities are "myths" (Fox, 2014, pp. 266–278). Not only are they exaggerated, he argues, but they are intended to elevate fads over facts. Even the risk of nuclear accidents, according to Fox, is remote. He provides careful analysis of past nuclear accidents, concluding that they occurred because the reactors had one-off technical problems that arose only because the technology was outdated. With more recent improvements in technology, he argues, the scientific evidence is clear that the chance of nuclear accidents is likely to be remote. The cost of construction, Fox admits, is significantly higher than for other sources. However, he argues that it is justified by many advantages of nuclear energy (see Fox, 2014, pp. 101–115; see also Rhodes, 2018, pp. 326–343). Reactors have a long lifespan. They have far less need for regular maintenance and repairs. Nuclear energy, he contends, is far more reliable than solar and wind.

Critics of Murphy, Fox, and other pro-nuclear advocates abound. They can be found in the circles of the Western Left Consensus. They range from individuals to organizations such as Greenpeace (MacDowell,

2017, p. 4). A concern about nuclear energy starting a new arms race and, ultimately, a nuclear war is recurrent (MacDowell, 2017; McNeil, 2007; Shrader-Frechette, 2011). Yet military research (Freedman & Michaels, 2019; Kaplan, 2016) systematically shows that the possibility of a nuclear war is remote. Nations possess nuclear weapons not for war but for deterrence. Protection against imperial interventions has been one of the most viable military uses of nuclear power. This is even more peaceful than using nuclear energy for self-defence.

Countries such as Libya that have given up nuclear power have suffered imperial interventions, which they avoided when they had nuclear technology. In their authoritative account *The Evolution of Nuclear Strategy*, Lawrence Freedman and Jeffrey Michaels (2019) note that "to the extent that there has been an effective nuclear strategy thus far it has depended on non-use, by deterring major war and helping to hold together alliances ... there has been no use of nuclear weapons since August 1945" (p. xiv). Of course, that does not mean there cannot be a nuclear war, but current evidence of the growing number of countries with nuclear capacity, sometimes called a "nuclear renaissance" (MacDowell, 2017), along with the proliferation of the idea of "minimum deterrence," gives grounds for some optimism. For example, the experience of China suggests that the risk of war is remote. China holds just enough nuclear capacity to deter others from unwanted imperial interventions in Chinese society. This "minimum deterrence" (Kaplan, 2016, p. 22) casts doubts on the suggestion of imminent nuclear war. As Kaplan (2016) puts it, "The reason this hasn't happened already is simple: the military, powerful factions of which are wedded to nuclear weapons, and Congress, powerful members of which have nuclear manufacturers or labs in their districts, won't allow it" (p. 22). In any case, the thirst for war is normally driven by business opportunities and the desire to expand private property rights in land. War itself creates opportunities, too, of course (Obeng-Odoom, 2019d), but claiming that they would be waged merely because of democratizing access to nuclear power for community, non-commercial uses in the Global South seems to be a stretch.

That said, more fundamental issues about nuclear power cannot be dismissed. For instance, Fox's case is also growthist – much like the Western Left Consensus. Also much like the Western Left Consensus, Fox's focus is disproportionately on the Global North, although he makes important observations about the primacy of Congolese uranium to the success of nuclear energy development (Fox, 2014, pp. 241–280). The dismissal of health concerns in nuclear communities is clearly hasty. Long-term ethnographic research, along with systematic historical work, compiled by Laurel Sefton MacDowell (2017) shows that these health issues are not mere speculation. There are serious health questions about nuclear energy development. The so-called success of France in recycling its nuclear waste (Fox, 2014, p. 4) is important to note, but it must not be disconnected from what Francophone Africa, especially the ex-colonies of France (e.g., Niger and Gabon) do for this imperium. They shoulder the burden of the initial and ongoing social costs of mining uranium for its nuclear energy program (Shrader-Frechette, 2011, pp. 215–217), while France monopolizes the long-term benefits of nuclear energy.

These problems with the Conventional Wisdom, however, do not imply acceptance of the Western Left Consensus. By neglecting colonially inspired rentier urban and regional development and by overlooking the specific property relations that shape the political economy of Africa's natural resources, the Western Left Consensus underestimates the centrality of the minerals-energy complex (MEC) to the world system (Fine & Rustomjee, 1996). Calling for a universal boycott of fossil fuel and nuclear serves as a manifesto, and when this Western Left Consensus approach recognizes the power of social movement and civic engagement, as well as current uprising, it can look inspiring. Yet it tends to neglect the details of what the uprisings are about, the demands of the protesters, and how they shift with time. Assuming that these are socialist revolutions, this hardly inspires any concrete steps for the various parts of the world system and their relationships to one another and to MEC.

In Africa, the question about oil – indeed, about landed resources – is far more complex. It is certainly not just about ending fossil capitalism, resource scarcity, or unreliable world prices. It is not even about the limitations of markets – what is sometimes called *energy security*. These are important, of course, but the coal-oil-nuclear question is also about energy sovereignty. Historical injustices, continuing dependent development, and colonial-urban and regional development, which echo and are echoed in spatial MEC, can be addressed neither by an oil optimist model nor by an oil pessimist alternative. Consequently, "Africa uprising" (Branch & Mampilly, 2015) is mostly about bringing changes in property relations, both locally and globally. In this respect, Marxist-inspired nationalizations or ideas for nationalization (see, for example, Amin, 1990, 2014; Nwoke, 1984, 1986) cannot meet the demands of popular protests.

Instead, concrete alternatives that transform existing exclusionary property relations have greater popular and analytical purchase. Commoning land, including oil, coal, and nuclear power, is one such alternative. Doing so means using rent taxation to end what Henry George (1885) once called "the crime of poverty". Addressing the paradox of "progress and poverty" (George, 1935) is a related strength. In practical terms, that economic paradigm entails not the nationalization of land, but rather the socialization of privately appropriated rents. Land that is not yet privatized can no longer be taken out of the commons. These steps go past energy security. They lead to energy sovereignty.

Energy Sovereignty

The silence on rent in the discussion on the energy question in Africa is deafening. For Henry George (1879/2006), however, analysing rent – how it arises and is shared, the opportunity cost of not privatizing the commons (i.e., not generating rent), the existence of rent and how it impacts ecological concerns – is central to the investigations for the keys to social progress, economic prosperity, and environmental sustainability. As chapter 3 suggests, rent arises in land, that is, in all natural resources, when the commons are commodified. It increases with social, public, and private investment, but it is appropriated by private landowners. The position of the rentier class, in terms of the rent it extracts, increases many-fold with population growth and speculation. Such increases, according to Henry George, are expressed in higher land and estate values.

In Africa, the economic rent in resource extraction is captured by oil companies. As rent is socially provided but only privately appropriated, the income divide widens between the expropriators of land and the rest of the society (Carmody, 2011; Obeng-Odoom, 2014b, 2020). Without taxing land – an interim measure to change resulting dynamics – or turning land back to common property, social problems will metastasize into a socioecological crisis. With unbridled privatization of land and the generation of more rent, the pressure to develop land further afield increases, as does rent in the core.

Georgism has recently made major inroads in China and elsewhere in Asia (Cui, 2011), but does it have anything to offer the political economy of Africa's oil resources? I will argue so. While there are not many Georgist analyses of oil in Africa, a few studies (e.g., Apter, 2005; Obeng-Odoom, 2014b; Olaniyi, 2008) point to directions that can be salvaged for such analysis. One relates to inequality between oil companies and the African countries where they operate. This maldistribution of rent arises from (1) oil contracts that are heavily skewed in favour of oil companies and (2) fiscal regimes that do not tax rents.

Another type of inequality relates to inequality between landlords, local people, and local oil communities. This arises from increases in site values, housing, and hotel prices, which are driven by migration into oil towns and cities, public investment in such settlements, and private investment, speculation, and expectation of prosperity. I have analysed many other types of inequality related to rent in Sekondi-Takoradi, Ghana. The extent of such experiences depends on whether the city abuts the oil field, is within the region of the field, or is the national capital. I document this process in *Oiling the Urban Economy: Land, Labour, Capital and the State in Sekondi-Takoradi, Ghana* (Obeng-Odoom, 2014b). The levels of inequality differ, but they share common features, such as the encouragement to exploit without concern for the environment and the absence of funds devoted to compensate those struggling from rent-related environmental crisis, which aggravate these contradictions.

Rent-related inequality – as with other types of inequality – leads to exclusion and marginalization, harsh accommodation conditions, and widespread evictions. In addition, the power to extract rent or the freedom from not paying rent leads the oil companies to extend and defend their property rights, at the expense of nature. Spillage aside, extractive activities disturb marine life, reducing fish catch for fishers in oil communities. In Ghana, for example, incomes of fishers and the size of the fish harvest have both declined as a result of extraction activities. While there is much talk about corporate social responsibility (Hilson, 2014; Obeng-Odoom, 2020), unequal power, arising largely from the creation and control of rent or appropriating rent, undermines attempts at democratically resolving the social, economic, and ecological crisis related to the extraction of oil.

George proposed two remedies for such problems, one interim, and the other more permanent. George's interim measure is to introduce land taxation, that is, require resource extractors to pay tax on their economic rent, super normal profit, or windfall. A tax should also be placed on the site value they (and others) appropriate. The revenues from this tax base can then be put into social investment. This taxation system has the additional benefit of reducing rising land and housing values and can divert some profits from oil companies to fishing and farming communities whose activities are disturbed by oil extraction.

While potent and used to success in Alaska, for example (see Widerquist & Howard, 2012a, 2012b), colonial and neocolonial processes such as neoliberalism have weakened state capacity in Africa so much that most African states do not even have the capacity to collect taxes. A UN-HABITAT report (UN-HABITAT, 2001) showed local revenues collected by city authorities in Africa is 11 times lower than the experience of industrialized countries. Even accounting for differences in land values, the state in Africa has much room for improvement. In the meantime, it is worth attempting to start collection now, even if only a small part can be realized. In future, however, George's proposed solution is to return land to its status as "common property."

This is where the Africans have greater leverage. While under the aegis of the World Bank, the IMF, and the German Development Bank, common land is being dissipated; customary land in Africa remains widely recognized and widespread. Colonial forces made the management of some of these lands undemocratic either by reifying fluid customs or privileging men's rights over women's, so some mangled "customs" ought to be revised. But contrary to the now popular view that the only way for Africans to "develop" is to create rent by privatizing and commodifying the commons, there is some research (e.g., Head-König, 2019; Obeng-Odoom, 2014b, 2020) that shows that using the oil commons, as suggested by Henry George, can bring progress without poverty or with rapidly declining poverty levels.

Implementing a Georgist philosophy, even his interim proposal of land taxation, is a Herculean task. For instance, attempts in Ghana to introduce windfall taxes at 10 per cent in the mining sector were blocked by mining interests. According to Ghana's former president, his government

> introduced a windfall tax, which is applied in several countries the mining companies come from for example in Australia and yet they will not allow us to implement a windfall tax in our country. They threatened to lay off workers if we implemented the windfall tax and because we needed the jobs and you don't want workers laid off you are coerced to go along. So, these are major issues we have. ("Windfall Tax Dropped," 2014)

While the 10 per cent tax rate is quite new, suggestions that the government was attempting something novel and was springing surprises on the extractive industries are unfounded. The country had a windfall profit tax in the 1986 Mining Code, pegged at a much higher 25 per cent, but it was reviewed and revised in 2006; by 2010 when the draft National Mining Policy of Ghana was introduced, the requirement for a windfall profit tax had been dropped (IEA, 2011). In 2012, its reintroduction was proposed, but the mining lobby defeated it. The spirit of George, however, would not lie still. The imperative for a mining tax emerged again in 2013 and in 2014, and again the mining lobby teamed up against it. This is just one example; the many perils of commoning nature should be carefully analysed.

Conclusion

The limitations of the Conventional Wisdom that advocates private TNC appropriation of the oil, coal, and nuclear commons are quite clear, as are the problems with the Western Left Consensus advocacy for renewables. The tragedy of privatizing nature is evident in the many social and ecological costs analysed in the chapter, but dealing with these problems by seeking to foreclose access to the "gifts of nature," as the Western Left Consensus advocates, deifies ecological problems that arise from privatizing nature. Renewable farms and parks benefit private TNCs. They stir multiple levels of dispossession (local people lose their land, common rents are privatized, and alternative uses of land are curtailed). Not only do these forces transfer socially created rent to absentee private landlords, but they also create local hierarchies. Wind farms give windfalls to landlords. For most people stuck in economic hardship and financial tunnels, they cannot see what *The Economist* ("A Dangerous Gap," 2020) calls the "rays of hope" from sunlight. Instead of renewal, their lives are trapped in blackouts too thick to be lifted by wind or illuminated by the sun.

This chapter has tried to develop a Radical Alternative, which acknowledges that oil-, coal-, and nuclear-related challenges and obstacles to mass prosperity are real. So are the limitations of placing faith in renewables as the new panacea. However, through the redesign of particular institutions, such as the socialization of oil rent and the pursuit of energy sovereignty, oil, coal, and nuclear power might be made a common property. The visible arms of the state could play a role in careful economic planning and development management, along with communities whose interests in these resources are non-commercial. Political elites and comprador state syndrome could be threats, of course, but there are other institutions (e.g., the unions, the media, and civil society groups working with communities and the state), in-state institutions such as the courts, and traditional institutions that could exert pressure for inclusive social change. External threats and obstacles faced by African countries would need to be dealt with by trying to overcome internal nation-centric strategies by developing regional, indeed multi-scalar, energy strategies.

None of these is going to be easy. Indeed, one oil minister, Sheik Ahmed Yamani, recognizing the complex political economy of oil, is reported to have said, "All in all, I wish we had discovered water" (Goodman & Worth, 2008, p. 201). Yet, in practice, the struggles around the water commons are probably even more complex.

Chapter Seven

Water

Introduction

The water commons in Africa is being rapidly privatized and marketized. However, unlike oil, this process is not well understood. Conventional Wisdom (for a detailed review of such analyses and arguments, see, for example, Connell & Grafton, 2011; Moyo & Liebenberg, 2015; Munck et al., 2015; Pearson & Kostakidis-Lianos, 2004; Sharma, 2012) claims that water markets have arisen because of the inferior nature of Indigenous systems. These customary alternatives are deemed incapable of offering precisely what water markets offer Africa: economic and ecological fortunes in a transparent system of water governance. From this perspective, markets are natural. They arise autonomously. They prevail over all other forms of governance. In this social Darwinist account of history, only the fittest and most deserving survives. So the marketization of water must be celebrated. Private property rights in water, these advocates argue (e.g., "The Climate Issue," 2019; "Markets in an Age of Anxiety," 2019, pp. 27, 28, 33; Turpie et al., 2008) – much like the propitious effects of clear private property rights on business activities (Redford, 2020), will make water resources in Africa more secure. Markets also will ensure better use of water, promote rapid economic growth and stir fresh ecological development.

In other versions of the argument for marketizing the commons, even if water commons do not evolve into private enclaves, they should be aided to do so (Alchian & Demsetz, 1973). For this purpose, T. L. Anderson and G. D. Libecap (2014, p. 204) argue that the conditions to be satisfied for the marketization of the commons are "well-defined, enforced, and transferable property rights." Some evidence (e.g., Boyd, 2012; Jeffords, 2015; Jeffords & Minkler, 2014; Obeng-Odoom, 2021) shows that when states make the right to the commons constitutionally enforceable, they actually help to protect the environment, new institutional economists (see, for example, Anderson & Libecap, 2014; Arezki et al., 2015; North, 1991; Redford, 2020) argue that problems related to governing the commons can arise only from the lack of markets or from the intervention of the state.

That is to say, only markets, to be precise "liquid markets" ("Markets in an Age of Anxiety," 2019, p. 33), are efficient. Working autonomously of social relations, markets are assumed to avoid the problems that arise from such relations. They produce better outcomes, too. Some new institutional economics (e.g., Ostrom, 1990) disputes Garrett Hardin's (1968) notion of the "tragedy of the commons" that claims that the commons are doomed to failure.

Yet even in this presumably milder version of the argument, there is an implicit endorsement of induced marketization. The promotion of private property rights *within* the commons is one way of doing so. In this path, the commons are merely the aggregation of joined-up *individual rights* called "joint" rather than common and equal rights (see chapter 2). This mechanism tries to naturalize the social basis of the commons (Euler, 2016). The idea is that without clearly defined individual property rights in the commons, the commons become mired in inefficiencies or remain "dead capital" (de Soto, 2000). Thus, although variations exist within the Conventional Wisdom, this body of knowledge is unified in its underlying logic.

Advocates of the Western Left Consensus contest the claims of the advocates of water marketization (for summaries of the debates, see, for example Duchrow & Hinkelammert, 2004; Manning, 2018; Obeng-Odoom, 2013d, 2021). They contend, instead, that the rise of markets is directed or imposed; that marketizing the commons is a path to social crisis; and that making access to the commons, especially the water commons, collective is more auspicious.

The unresolved questions in this debate are the following: (1) Were there markets in the beginning? If so, how have they transformed and, if not, how did markets arise and evolve over the years? (2) What are the outcomes of such markets for people and environment? (3) How should we interpret the outcomes of water markets and should water be commodified at all? Such questions are core to the historical debate between new institutional economists and heterodox or original institutional economists (Bromley, 2019; Obeng-Odoom, 2015a). The existing studies are not formulated to address these key questions, or when they do, they are mostly theoretical in nature. For instance, Jeffords and Shah's (2013) important work in *Review of Social Economy* uses a rights-based approach to analyse the commodification of water, but the interest is mainly theoretical and the emphasis is on expanding the neoclassical models on water, not on addressing the historical questions between non-neoclassical new institutional economics and heterodox original institutional economics research. Empirically, far more effort has gone into research on "land grabs" (see special issues published, among others, in *Journal of Peasant Studies*, volume 39; also see the review by Renom et al., 2020).

Thus, in an important contribution to *Marine Policy* journal, N. J. Bennett and his colleagues from the Institute for Resources, Environment and Sustainability at the University of British Columbia and the World Commission on Protected Areas and Locally Managed Marine Area Network in Suva, Fiji, called for "empirical case studies that document ocean grabbing in different locations" (Bennett et al., 2015, p. 65). The aim of this chapter is to address the three key questions in the commons debate. I use empirical examples from Africa, especially Ghana, where the marketization of water is ongoing at a pace never before witnessed in Africa's "geographies of change" (Grant, 2015). How should water markets be analysed? A Radical Alternative must be holistic, not simply mechanistic or ethnographic, as recent work (Renom et al., 2020) recommends. Such methodological holism must entail the following:

(1) The question about the rise of water private property rights (q1) requires first an analysis of the historical roots of the context under discussion. Second, the approach calls for a detailed examination of marketization: a political-economic concept that "denotes the expansion of market coordination into non-market coordinated social domains as well as its intensification in already market-dominated settings" (Ebner, 2015, p. 369). Marketization can also refer to the realignment of markets to serve even greater capitalist processes (Stilwell, 2011b). Being historical, this institutional political economy framework can help to better understand not only what the current marketization situation is but also how and why it came to be and in what ways it has been transforming.

Unlike the neoclassical economists who calculate causation, the institutionalists consider context and embrace competing explanations using contextual tests of probability, when appropriate, to determine causation or multiple causations (Lenger, 2019;

Figure 7.1. Criteria for Judging Outcomes in the Debate between Competing Institutionalist Paradigms.

(d) Opportunity cost	(a) Absolute outcomes
(c) Congruence between promises and outcomes	(b) Relative outcomes

Morck & Yeung, 2011). The historical emphasis in institutionalism is also unlike the new institutional economics variant of neoclassical economics, which is historical, too, but is short to medium term in its historical analysis; institutional political economy draws simultaneously on short, medium, and long-range historical accounts, emphasizing the plurality of time and different social contexts. More so, this institutional political economy is evolutionary, and the methodology is one of continuous evolution under a wide range of influences; hence, local histories are linked to regional, international, and global histories (Obeng-Odoom, 2016a, 2016b, 2021). In short, this chapter combines historiographical probes with world-systems analyses, inspired, among others, by the work of Immanuel Wallerstein and Fernand Braudel (see Lee, 2012, for an overview), to answer question 1.

(2) For question 2, that is, establishing the outcomes of commodification (q2), at least four types of analysis are done, as shown in Figure 7.1.

The first criterion, (a), evaluates absolute (dis)advantage defined here as the "bare" outcomes of commodification (e.g., generating jobs regardless of how many or the quality of the jobs). The second, (b), determines success or failure by relating those outcomes to the quantum of resources transferred into private control. The third, (c), measures success or failure based on whether there is congruence between promises made by private interests and the practices of such private interests. The fourth, (d), assesses the outcomes in (a) to (c) by reference to alternatives that pre-existed or are fundamentally different from the current commodified water regime. This four-quadrant decision rule or approach to analysis is modest, but it is substantially wider than approaches that are used in existing studies either of water (e.g., Anderson & Libecap, 2014; Libecap, 2018; Schoneveld et al., 2011) or land (Arezki et al., 2015; see also papers in *Journal of Peasant Studies*, volume 42, issues 3–4) that centre on the criterion in quadrant (a) or (b). The approach for this chapter, on the other hand, provides a much wider framework of analysis for that purpose.

(3) On how to interpret the outcomes of water markets and whether water should be commodified at all (q3), we need to look at a so-called free-standing "economy," but we need to look additionally at how it is embedded in society and nature (Polanyi, 1944/2001), as well as at its social and ethical foundations and ramifications (Gonce, 1996). Efficiency is not simply an economic but also a social, ethical, and culturally sensitive question (Hill, 1966) – a useful approach in a world in which different societies and diverse groups in societies have varied and variegated ways of life that evolve over time under different influences. Consequently, this approach avoids the "conceptual bias" in neoclassical economics (Elahi & Stilwell, 2013) and its new institutional economics variants (for the new institutional economics methodologies, see, for example, Alchian & Demsetz, 1973, and North, 1991). It also reviews the different methodologies, such as Claude Ménard and Mary Shirley's (2014). And it is politically more explicit in its pursuit of justice than some new institutional economics work, such as Elinor Ostrom's (1990), that delves into (property) rights without embracing the politics and values of social justice and naturalizes the commons without taking into account their social underpinnings and ramifications for society (Euler, 2016; Exner, 2015).

(4) The empirical referent is Ghana. It is a useful case because the process of privatizing water there has been unfolding for some time now, but much of the research has centred on the commodification of land (e.g., Asante, 2020; Boamah, 2014a, 2014b), which is important to understand the wider process of commodifying the commons but has to be complemented with analysis on inland water systems, particularly at a time when the country is seeking input into the preparation of its 40-year development plan.

Overall, the empirical evidence shows that markets have been socially created. Through imposed and directed efforts, these processes go beyond the recent turn to neoliberalism (the emphasis in the Western Left Consensus). Some jobs have been created through investment, but such employment is not unique to marketization and private investment. Indeed, the private model of property rights has badly worsened the distribution of water resources, not only within different property relations in Africa but also between diverse property relations. Water markets have been responsible for much displacement and trouble for communities and for nature. Overall, there is no necessary congruence between the promises made by advocates of the Conventional Wisdom and how communities experience water markets. In contrast to the Western Left Consensus and its causal theory of neoliberalism, the commodification of water has a much longer history. Indeed, tighter state regulations for the use of inland and transboundary water sources might temporarily halt the displacement of communities sparked by marketization of the commons, but only one fundamental change can guarantee community well-being. Regarding access to and community control of water as constitutionally sanctioned human rights and as *res communis* collectively constitute that Radical Alternative.

The rest of this chapter is divided into three sections, each devoted to one research question. The next section looks at the origins of water markets, looking at (1) the pre-colonial communal water property rights system, (2) the colonial and postcolonial reforms, and (3) the neoliberal experience. Section two examines the outcomes of such markets for society, economy, and environment; and section three seeks to interpret the outcomes of water markets and the marketization of the commons.

The Origin of Water Markets

Attempts to answer such questions by looking at regional trends has neither succeeded in providing needed explanations nor given any clarity to the issues (e.g., Gellers, 2012, 2015; Kwoyiga, 2019; Manning, 2018). In the case of Ghana, at least, it is more effective to do a local historical analysis combined with global-systems thinking within three-time frames. First is governance of water in the pre-colonial era. Second is the colonial and colonizing water reforms. The third is an articulated analysis of postcolonial neoliberal reforms in the twenty-first century – and how they intersect and complement one another. For the pre-colonial situation, two key sources of history are S. H. Hymer's "Economic Forms in Pre-colonial Ghana" (1970) and H. D. Dyasi's "Culture and the Environment in Ghana" (1985). These accounts require some attention because they contain original historical evidence of how water was governed.

According to Hymer (1970, p. 34), "In the pre-colonial Ghanaian economy, land was distributed equally and most families had full rights to the land they used, paying little or nothing in the way of rent or taxes." He continues, "A fair share of available land was the right of every member of the community. The products of the land thus belonged to the family that cultivated it and there was no leisure class deriving its income from rents." It is not that the economy was fully egalitarian: a leisure class existed but it extracted its rents from trade, mainly. The low- or no-rent economy did not have a basis in the sheer fact that land was abundant relative to the population. Rather, as Sackeyfio-Lenoch (2014) shows, the economy was based on Indigenous principles of sharing the commons which were *excludable*. People outside the landowning community could not freely access land. Similarly, strangers living in the community had restricted access to land but not because land was scarce.

Indeed, the idea of "scarcity," especially market or artificially created scarcity, was actively opposed by the Indigenous system, which was based on the principle of sharing. Restricted access to land was simply on grounds that people were not members of a certain landowning group. The chiefs, priests, or family elders that led the landowning communities were not landlords in the European sense, but trustees. Households were quite autonomous, living on what they produced, some of which was exchanged with other households and, on a limited level, with other communities. So, markets existed but not in land, not for profit-making purposes, and not autonomously. In short, these markets were socially moulded. Much like what is described by Karl Polanyi (1944/2001), markets are socially constructed and are embedded in social relations, and once they remain, there is social, economic, and ecological harmony. In contrast, if the market over-reaches its social boundaries – or attempts to do so – society stands dislocated, causing a backlash of social ills and then a challenge to the ongoing process of marketization. That is what Polanyi (1944/2001) called a "double movement."

Customarily, land and water are analogous. Indeed, land is water (Ollenu, 1971, p. 135; see also chapter 1). Property rights in water were traditional and communal (Gyau-Boakye, 2001), so "water property rights" in this context had an entirely different meaning to the same phrase used for those with private rights over inland water sources. Water property rights were held in trust and regulated by traditional leaders. Communities had common property rights to water. Based on riparian rights or prior-appropriation rights, people individually and collectively used water resources. In this commons, economy, society, and environment were mutually supportive, mutually nourishing, and

Figure 7.2. Indigenous Web of Life.

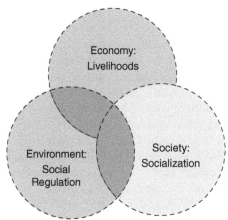

mutually empowering (Figure 7.2). In turn, the Indigenous system guaranteed certain standards of livelihoods, environment, and socialization without destroying their intersectional elements.

In terms of socialization, water was sacred – much like land and indeed the earth – and was a nucleus around which the society solidarized, united, and reflected on their existence. Women and men both used water albeit for quite different but related reasons. Women and children used water to cook, to take care of the household's feeding and drinking needs, and to fulfill domestic fish requirements. Men used water for farming, livestock rearing, and home/exchange fishing purposes – as G. K. Nukunya's *Tradition and Change in Ghana* (2003, pp. 94–106) shows. Water festivals were held to honour water bodies and their gods. Such festivities were also periods to ponder issues facing the communities, which could be economic, environmental, or social. These interconnections between society, economy, and environment were a core culture of pre-colonial Ghanaian communities. "Among the Akan," Dyasi (1985, p. 99) has noted, "bodies of water are associated with gods or *abosom* and must be used in accordance with their wishes." Economic and social activities were clearly socially regulated to prevent disease and pollution. According to Dyasi (1985, p. 99), "Those parts of the river used for bathing, swimming, or for watering domestic animals are located downstream in relation to those used as sources of drinking water … dwelling places are often situated more than half a mile from rivers … water conservation is a common tradition." Farming around such bodies was controlled by

Indigenous law, beliefs, and practices, such as maintaining a clean environment around the water bodies where gods were believed to reside, and fishing was prohibited on certain days when the gods are resting (Entsua-Mensah, 2001).

Water resources were used for livelihood activities such as farming, herding, and fishing. These activities were restricted to community members. Institutions such as that of chief fisher or the *apofohen* in Akan-speaking areas ensured the stability of this system of water communal ownership and use. While hereditary, chief fishers had to be elected by other fishers for the position to be effective. The *apofohen* is supported by a committee or counsel of Elders (*beesonfo*) with whom they carry out their duties. Their responsibilities include solving conflict arising over the use of water and providing laws on which types of fish to harvest and at what time. Regulating days to fish and which fishing gear to use, the chiefs of the sea command significant respect and, together with the priests of the sea, offer prayers to traditional gods in charge of water bodies and fisheries (*bosompo*).

Depending on how important fishing is to any community, these chiefs of water are very powerful and enforce territorial water rights (Overå, 2001). They and others regulate fishing as an economic activity. For some communities, fishing is banned in some river bodies. That is evidently the case in Berekum where *Asuo Koraa* or River Koraa is regarded as female with many children (fish) and, hence, traditionally, no fishing is permitted in the water. In other rivers in Berekum, chemical use for fishing is banned. Among other groups, fishing is not allowed on *nnabɔne* or taboo days (Awuah-Nyamekye, 2013).

Such practices also helped in maintaining the forests and biodiversity of flora and fauna, in the sea and on the land. Some estimates put the biodiversity in the country around 2100 plants species and at least 200 species of mammals, 74 species of bats, 37 species of rodents, and 200 species of birds in the forest zones of the country (Francois, 1995, p. 122). The biodiversity differed according to different vegetation type across the country but "irrespective of which vegetation type they were in, the communities had formed an enduring association with their environment which impacted on their material, cultural and spiritual life" (Francois, 1995, p. 122). The people "recognized the forest as an important renewable resource which satisfied almost all their basic needs and they used it wisely ensuring that the resources of the future were not depleted" (Francois, 1995, p. 123). J. H. Francois (1995) has noted that "the forests were communally owned, managed and utilized and in doing this successfully our ancestors were aided by low populations, low demands and a relative abundance of resources, sustained by using the accumulated wisdom of the ages, residing in the elders as custodians" (p. 123).

Abuses of the system were checked by Indigenous processes and institutions. Systematic studies – such as Sara Berry's (2001) *Chiefs Know Their Boundaries: Essays on Property, Power, and the Past in Asante, 1896–1996*, Arhin Brempong's (2007) *Transformations in Traditional Rule in Ghana (1951–1996)*, and Naaborko Sackeyfio-Lenoch's (2014) *The Politics of Chieftaincy: Authority and Property in Colonial Ghana, 1920–1950* – show that prior to colonial domination, chiefs were more accountable to their people and decisions were more collectively made. The youth (through organizations such as the *asafo*) were active in holding chiefs to account. Chiefs whose actions (e.g., selling community landed resources) were not congruent with community aspirations were usually removed from their position.

Water, forest, and animals were communally managed. The purpose was not only to conserve but also to expand the ecosystem. The vision was not solely because of a reverence for future generations but also because of a respect for nature. Nature was believed to possess spiritual powers (Eshun, 2011). Research by Walker (2002) shows that these commons were excludable. Indeed, the traditional leaders responsible for water management in certain areas took some payment from migrant fishers who visited their territories to fish. The payment was not always money: it could be fish or drinks (Entsua-Mensah, 2001). Traditional priests and other traditional leaders played a similar role with respect to other water bodies such as rivers, lakes, lagoons, and reservoirs. Existing studies (e.g., Awuah-Nyamekye, 2013; Dyasi, 1985; Hymer, 1970; Kwoyiga, 2019; Tsamenyi, 2013) suggest that, on balance, the traditional system was successful in providing security of tenure to water resources and in supporting the economic, social, and ecological needs of local people.

Colonial and Postcolonial Reforms, 1844–1970s

The institution of the central colonial state in 1844 transformed property rights in Ghanaian water bodies. They became delinked from economy and society in the ways that they been before, nourishing and expanding the social foundations of communities. They became statist and bureaucratically managed. Property rights became commodified in ways that transferred/transformed community benefits to private individuals and corporate entities seeking to profit from common resources.

As noted in the stimulating book *Decolonizing Nature* (Adams & Mulligan, 2003, pp. 1–15), British imperial policies about conservation were underpinned by ideas that were fundamentally at odds with pre-colonial forms. Such policies conceived of "nature" as something "out there" to be "conserved" and "protected" from humans. Sharp and piercing, these policies slashed the interrelationships among economy, society, and environment. British policies often prioritized the economy over and above the rest. Modern technology and bureaucratic top-down laws could not comprehend the complexities of common property rights. The egalitarian underpinnings were replaced with ideas of strict hierarchies between nature and humans. This British-imposed system was patterned after the hierarchies between the colonizer and the colonized.

The colonizer saw nature as something to be subjugated for profit using "superior" technology and "experts." British Indirect Rule in Ghana was complex in its obtuseness, subtlety, and co-option. This rule was more about being seen to be supporting Indigenous forms than actually doing so. The end product of all such "support" was to elevate the interest of the colonizer over that of the colonized (Obeng-Odoom, 2014c; see also Manning, 2018) whose animistic practices were deemed "inferior," "fetish," "superstitious," and "backward" (Eshun, 2011; see also Manning, 2018). With animism, sharing, and community undermined, commercialization was extolled to the place of religion. Racial supremacy was institutionalized (see Obeng-Odoom, 2014c; see also Manning, 2018) and propelled by imperial deceit and gun power. To crush local protests and resistance or what Francois (1995, p. 124) called "indigenous militancy," a local arms race was set in motion. In turn, the grounds were cleared for dramatic changes in the governance of common water resources.

It appears that the first legislation about the commons – water and fishing, to be more precise – was passed by the colonizer in 1903 (Dyasi, 1985). Yet imperial judicial pronouncements existed long before then. Whether in the form of legislation or judicial precedent, however, the colonizer provided simplistic European views inconsistent with the complexities of the commons in Indigenous society (Walker, 2002). In the well-known 1898 case of *Akwufio and Others v. Mensah and Others*, heard in the British created Supreme Court, Akwufio and colleagues had flouted a local law passed by Mensah, a traditional water/fishing chief. The law was against the use of a particular piece of modern fishing equipment, the *Ali* net, that had been found to increase fish harvest but at the cost of creating overfishing and driving down fish stocks. When Akwufio went against the law, he and his team were reprimanded.

Dissatisfied, Akwufio and colleagues took the case to the imperial judge, Sir Griffith, who had no problem upholding the case of modernization. He went as far as claiming that it was quite common for the *Ali* net to be used in England, so it must be a good thing. In turn, Sir

Griffith undermined the traditional authority of Mensah, while extolling Western values and ethos as superior (Walker, 2002). A legal commentator on the case (Walker, 2002) has pointed out that Sir Griffith's ruling reflected a general perception held by the Europeans of the time that African waters were "no man's land," they were in abundance, and they did not have specific property rights. This interpretation is consistent across several water/fishing-related cases. In 1906, Mr. Elliot, a colonial officer appointed to mediate a conflict between Ghanaian fishers, explicitly stated that water property in Ghana was common and "open access" (Walker, 2002, p. 398), implying it was "for nobody," open for accumulation and profiteering. Later, in the 1934 Winneba Case, a chief's ban of modern but ecologically harmful fishing practice was reversed by the colonial secretary of agriculture who claimed that "the best fishing net is the net which catches the most fish" (as cited in Walker, 2002, p. 397). This productivist logic persisted throughout the colonial era.

In 1946, for example, the colonizer founded the Department of Fisheries to oversee processes to increase the fish harvest. The postcolonial state, in spite of its revolutionary orientation, inherited the "modernist" approach. So in 1964, the postcolonial government provided "modern" methods of increasing the fish catch. While some local chiefs in charge of water resource management intervened by making local laws about which fishing methods to use for sustainable fishing, the Western court structures delivering "British customary law" did not support the Indigenous processes (Tsamenyi, 2013). With time, overfishing became a major policy concern, and once again modernist, Western-funded project reforms were recommended to the government (Tsamenyi, 2013). A distinctive feature of this new phase of Westernization was the more explicit enclosure of water property rights. However, like the previous attempts at creating property in Indigenous resources through undermining Indigenous institutions in the name of "modernization," the end was to obtain economic efficiency (Tsamenyi, 2013).

British colonialism, and with it greater marketization of the commons, also systematically undermined accountability. For example, through the delegitimization of the authority of the *asafo* movement, the excessive concentration of power in the class of chiefs, and growing opportunities to make money, the abuse of chiefly powers became more common as did the "sale" of land (and water) to strangers (Berry, 2001; Brempong, 2007; Obeng-Odoom, 2020, 2021; Sackeyfio-Lenoch, 2014; Ubink & Quan, 2008). This reconfiguration of land (and water) governance was a calculated attempt by imperial Britain to administer the country in ways congruent with British commercial interests (Berry,

2001; Gyampo & Obeng-Odoom, 2013; Nyantakyi-Frimpong, 2020). That system was inherited by postcolonial regimes.

Still, the post-independence government of Kwame Nkrumah made some crucial changes to the water sector. Among them, the promulgation of the Volta River Development Act of 1961, which provided the legal basis for the construction of a hydro dam on the River Volta. Four years later, Nkrumah's government passed the Ghana Water and Sewerage Corporation Act that set up a major semi-autonomous state institution: the Ghana Water and Sewerage Corporation. The organization was mandated to ensure that drinking water in the country was potable. After the fall of the government of Nkrumah, others promulgated important legislations. Notable among these was the establishment of the Council for Scientific and Industrial Research in 1968 with research cluster devoted to water research (Dyasi, 1985). Laube (2005) has offered a detailed review of specific laws about water promulgated in the post-independence era, while Tsikata (2006), Owusu-Ofori and Obeng-Odoom (2015), and, more recently, Amankwah-Amoah and Osabutey (2018) have analysed the impacts of establishing the 1961 dam on some nearby local communities. The rest of this chapter is not a vague and generalized commentary on neoliberalism, the stock-in-trade of most writing on neoliberalism (Dunn, 2017), but, rather, is a careful analysis of sweeping neoliberalism in water policymaking.

Sweeping Neoliberal Policy Reforms

Since the 1980s, neoliberalism – a political-economic concept, a set of ideas, or even a framework used to describe the commodification of all sectors of society (Cahill et al., 2012, p. 6) – has become the ruling ideology of the Ghanaian state, regardless of which political party is in office (Bob-Milliar, 2019). R. H. Green (1987) prepared a detailed country study on Ghana, emphasizing the nature and effects of the stabilization and adjustment programs in the country. However, Green (1987) did not examine the marketization of the commons, so this section of the chapter fills in the gap. The neoliberal orientation reflected internal experiences with so-called failed "socialist" experiments and international pressure as reflected in the structural adjustment programs (SAPs) imposed on the country as part of the Washington Consensus and the directed Poverty Reduction Strategy Papers that replaced SAPs (Obeng-Odoom, 2013b). Constitutionalism and market democracy were also part of the imposed package, recalling J. R. Commons's (1924) emphasis on "the legal foundations of capitalism." In turn, the then military-based Provisional National Defence Council (PNDC) government hurriedly transformed itself into a civilian party, the National Democratic Congress, to contest political elections (Oquaye, 1995a, 1995b), which the party won amid now well-known accusations of electoral fraud (Oquaye, 1995a, 1995b).

It was within this context that the current 1992 Constitution of Ghana was promulgated. The Constitution explicitly recognizes the position of chiefs regarding the fiduciary management of customary land rights. According to Article 267 (1) of the Constitution, "All stool lands in Ghana shall vest in the appropriate stool on behalf of, and in trust for the subjects of the stool in accordance with customary law and usage." The "stool" is the symbol of authority of chiefs in southern Ghana. The equivalent symbol of authority in Northern Ghana is the skin, to draw on explanation contained in the *National Land Policy* (Ministry of Lands and Forestry, 1999). Jurisprudentially, the root of title in both southern and northern Ghana is the community, with traditional leaders acting only as trustees (Asante, 1965; also see Ministry of Lands and Forestry, 1999). According to article 277 of the Ghana Constitution, "'chief' means a person, who, hailing from the appropriate family and lineage, has been validly nominated, elected or selected and enstooled, enskinned or installed as a chief or queen mother in accordance with the relevant customary law and usage.'" So, chiefs of water or simply the *apofohen* position is backed by constitutional provision (also see sections 57–58 of the Chieftaincy Act, Act 759). To emphasize a conceptual point developed in chapter 1, land is construed as including water. According to Justice N. A. Ollenu (1971), an authority on the subject and former chief justice of the Supreme Court of Ghana,

> The first concept of land tenure in Africa – Ghana and Nigeria in particular –
> is that absolute and unqualified ownership of land is vested in a commu-
> nity. This community ownership implies not just the right to the use and
> occupation of the land but also to the ownership of the soil, minerals and
> all things under the surface, rivers, streams and watercourses on, and the
> right and jurisdictional authority over, the land. (p. 135)

Therefore, by guaranteeing Indigenous land rights, and the position of their custodians, the communal and customary basis of water had to be guaranteed, too. One exception is when compulsory acquisition is appropriately publicized and compensation is paid to the expropriated people. Article 20 of the current Constitution – titled "Protection of Deprivation of Property" – is clear on the nature of compulsory acquisition in Ghana:

(1) No property of any description or interest in or right over any
     property shall be compulsorily taken possession of or acquired by

the State unless the following conditions are satisfied. (a) the taking of possession or acquisition if necessary in the interest of defence, public safety, public order, public morality, public health, town and country planning or the development or utilization of property in such a manner as to promote the public benefit; and (b) the necessity for the acquisition is clearly stated and is such as to provide reasonable justification for causing any hardship that may result to any person who has an interest in or right over the property.

(2) Compulsory acquisition of property by the State shall only be made under a law which makes provision for (a) the prompt payment of fair and adequate compensation; and (b) a right of access to the High Court by any person who has an interest in or right over the property whether direct or on appeal from other authority, for the determination of his interest or right and the amount of compensation to which he is entitled.

(3) Where a compulsory acquisition or possession of land effected by the State in accordance with clause (1) of this article involves displacement of any inhabitants, the State shall resettle the displaced inhabitants on suitable alternative land with due regard for their economic well-being and social and cultural values.

The year 1996 was a watershed in the marketization of the commons in Ghana. Specifically, that was the year in which the state purported to unilaterally extinguish customary rights in water (Williams et al., 2012). In that year the Water Resources Commission was established by the Water Resources Act (Act 522). The paramount aim of the Commission has been to create, extend, and sustain private property rights in water. It does so by issuing and maintaining a system of permits and licences to use and private ownership and control over water. Section 12 of Act 522 decrees that "the property in and control of all water resources is vested in the president on behalf of, and in trust for the people of Ghana." The Act is complemented by the LI 1692; the Water Use Regulations, 2001; the National Integrated Water Resources Management Plan; and the National Water Policy, 2007, to provide a legal and economic basis for the creation of private property rights in water as the state is empowered to parcel out shares in water in exchange for money.

The legal structure seems to be dealing with water usage rights, rather than rights to exclusive possession over an area of inland water. Indeed, according to section 1 of LI 1692, titled "Water Use Permit,"

Subject to the Act, a person may obtain a permit from the Commission for: (a) domestic water use (b) commercial water use, (c) municipal water

use, (d) industrial water use, (e) agricultural water use, (f) power generation water use, (g) water transportation water use, (h) fisheries (aquacul-
ture) water use (i) environmental water use (j) recreational water use, and
(k) under water (wood) harvesting.

However, according to section 25, interpretation, of the LI, "'use'
means any taking advantage of water, either as a physical substance
or as a water body to meet any demand" (p. 10). It follows that the
two – right of use and right to exclusive possession – while different
are, in essence, similar. They operate to limit access by local people
to both use and the physical resource because some water bodies are
located on land that is leased exclusively to some companies (Williams
et al., 2012) whose parcels can, within the law, be enclosed and encased
within walls or fences.

While the legislation seems to contain some faint traces of pre-existing
customary arrangements, this remnant, too, is a poor reflection of
customary practices. According to regulation 10 (1) sub-regulations
(a) and (b), some people are exempted from seeking permits but only
if they are involved in subsistence agriculture on land not bigger than
one hectare or, if using mechanical methods, a person does not abstract
more than five litres per second. Even in these cases, however, they
have to register their use with the local government – not with the traditional authority.

These reforms arguably provide the basis for a major agrarian change
in the country's ambition for an agro revolution. The draft Bioenergy
Policy for Ghana (Energy Commission, 2010) makes a strong case for
the promotion of commercial investment in land for the cultivation
of agrofuel crops. A program of Accelerated Agricultural Modernization and Commercialization for Increased Food Security and Economic
Transformation was launched in 2009. And, in the following year,
the then minister of food and agriculture, Kwesi Ahwoi, announced
in Washington that there was plenty of vacant land to be obtained in
Ghana and gave an open invitation to large capital and the powerful
states of the world to invest in biofuel and food cultivation in Ghana
(Ahwoi, 2010). As noted by Libecap (2018) and Anderson and Libecap
(2014), for this investment to be realized, private water property rights
regime is a prerequisite.

The existing legal regime in Ghana, which makes it possible to obtain
private property rights in water, therefore, successfully establishes the
locus standi to convert customary rights in water to state property and
then to private water property rights, often held by transnational agrofuel investors, among other TNCs. According to the then minister, as of

2010, over 20 companies from different parts of the world such as China, Brazil, Germany, Italy, The Netherlands, and Norway were already in the process of obtaining land in Ghana for biofuel crop production (e.g., jatropha and sugarcane) (Ahwoi, 2010, p. 14). Recent research on the topic (see, for example, Yaro & Tsikata, 2013) suggests that the number has increased. Examples of water markets have been offered by Kizito et al. (2013, pp. 342–343): the Solar Harvest in Yendi is into irrigation agriculture for which it is to draw water from the Bontanga Irrigation Scheme, with plans to pump water out of the White Volta River. Kimminic Estates Limited in Kobre has acquired 43,000 hectares of land from which the company pumps water for its agrarian activities.

The experience of Ghana ought to be placed in the wider context of the global push for the marketization of the commons, of the environment, and of water (Stilwell, 2011b; "Markets in an Age of Anxiety," 2019; Zhang, 2017). In the 1970s, precisely in 1977, the United Nations supported the organization of the conference Mar del Plata, which led to the declaration of a UN Water Decade (1981–1990). A follow-up world meeting took place in New Delhi in 1990, during which a collective, communal approach to water was asserted (Franco et al., 2013). It would not, however, take root, as the International Conference on Water and the Environment held in Dublin in 1992 articulated and strongly championed the opposite position: extolling the virtues of private property rights in water resources (Franco et al., 2013). In 1996, the Global Water Partnership and World Water Council were founded to better articulate and champion the view that water be regarded as an economic good and that greater privatization and corporate investment would lead to more efficient use of water resources.

The founding of the World Commission for Water in the twenty-first century followed subsequently and argued that better defined water rights are a prerequisite for pro-poor and transparent water use. Much of the work of the world development bodies, notably the World Bank but also the Swedish International Development Agency, subsequently focused more strongly on ways to better define water rights and make water a commodity (Bond, 2010). These developments are significant because they censored the non-pecuniary attributes of water and, instead, tried to solve the diamond-water paradox posed by Adam Smith. It did so by treating water as possessing economic value, like diamonds.

The eclipse of progressive discourse on water and the ascent of the new movement that placed primary emphasis on water as commanding "exchange value" was not an event. It was a process of several attacks on the commons. Intellectually, the right-wing Institute of Economic

Affairs published *Markets under the Sea: A Study of the Potential of Private Property Rights in the Seabed,* a book by D. R. Denman, notably the first full professor of land economy at Cambridge University. Denman argued that the state should assume ownership of water rights and then parcel out property rights in water to individual interests as the best way to attain efficiency. Hence, economic efficiency, of creating the highest rent, took precedence over the social and ecological efficiency of water.

The typical assumptions were that "common property" meant, to use Professor Wiseman's words, "no property" (Wiseman, 1984, p. xiii). That is, the sea was an open range system, and as such systems are often the subject of abuse and inefficient use, it is much better to create private rights under the sea, because nationalizing it, the only other option, was inequitable and similarly inefficient. The idea of *integrated water resource management* (IWRM) captured the vision: making water a commodity (Franco et al., 2013).

In Africa, the marketization of the commons was embraced in the Ouagadougou Ministerial Statement on water in 1998 after the West African Ministerial Conference on IWRM held in Burkina Faso from 3 to 5 March 1998. A Technical Committee and a Coordination Centre emerged out of the conference, with a joint mandate to urgently change how water is seen: from a natural resource with use value to an economic commodity with exchange value (Water Resources Commission, 2012). This transformation was effectively a continuation of the sweeping reforms in Africa in the 1980s and early 1990s under the rubric of SAPs, which were directed, imposed, or funded by the Bretton Woods Institutions, having in common a shared commodity vision for common and public resources, and how they can best be managed (Obeng-Odoom, 2013b).

International and continental agreements or statements have been written to provide a strong basis to legitimize and facilitate such marketization of the commons, while delegitimizing customary systems. This twin trend is part of a long-term process of using Africa as a laboratory for sweeping privatization programs funded, directed, or imposed by the Washington Consensus in the 1980s and early 1990s. Since then, other international and continental agreements have sustained and increased the interest in creating markets but, this time, in water. The 1992 Dublin Statement on making water a private, economic good is one example, as is the Ouagadougou Ministerial Statement on water issued in 1998. Together, these processes provide a strong legal basis to mark a *right* turn to water property rights systems (Water Resources Commission, 2012). Today, under the SDGs, the Paris Agreement, and a global political economy where "liquid markets" reign ("Markets in an Age of Anxiety," 2019, p. 33), water, a hitherto non-pecuniary natural resource, is widely regarded as a commodity, and transnational interests have identified Africa as a booming market where water rights can be bought and sold. These markets – whether called clean development mechanisms or framed as sustainable development mechanisms – are widespread.

They can be called "water grabs" (Allan et al., 2013). One example is the Lesotho Highlands Water Project, which enables South Africa to draw water from Lesotho to its Gauteng Province. Another example is they way Agri South Africa, the commercial farmers' union of South Africa, has literally arranged to buy large parts of inland water bodies in the Democratic Republic of Congo as part of 10 million hectares of farming land offered to the group by the Republic of Congo (Sebastian & Warner, 2014, p. 10). Other examples of the privatization of water bodies can be found in the Cameroon Development Corporation plantations in South Western Cameroon, where corporate banana plantations are developed in water-"abundant" areas that become ring-fenced for corporate water interests, leading Fonjong and Fokum (2015, p. 115) to argue that "hidden behind every land grab is a water grab."

This observation is valid for Ethiopia too. In *Land Grabbing: Journeys in the New Colonialism,* Stephano Liberti (2013) notes that, in Ethiopia, "as part of the land leasing policy, the use of water for irrigation is included in the price … The lands rented to Karuturi and Al Amoudi in the Gambella region are located on the banks of the area's main water tributaries, which are a primary source of life for the Indigenous population" (p. 38). In the same country, dams have been built on large rivers, such as Omo and "along the course of the Blue Niles and its tributaries," to create electricity and to irrigate new landed investments (Liberti, 2013, p. 37). So, water becomes a fundamental umbilical cord that holds together capitalist investments in nature. The question is not as simple as asking whether hydropower is a "saint or sinner," as Yan Zhang (2017, p. 99) contends. Indeed, it is not even about taking a more practical stance, the so-called balanced view between the Conventional Wisdom that extols the dam as the centrepiece of developmentalism and the Western Left Consensus that is fundamentally against dam projects (Zhang, 2017, pp. 114–115).

Rather, developing a radical view implies going beyond both positions. In this sense, it is necessary to investigate the creation of autonomous private water rights such as irrigation schemes, another commonplace strategy. The government of South Sudan plans to allow the government of Egypt to construct a canal to suck water to Egypt, that is, water from the Nile that is currently used by the giant Sudd Swamp in South Sudan (Pearce, 2012, p. 49). The Zimbabwe Bio Energy

Company is another case. A corporate entity made up of "former white commercial farmers" working with "an indigenous organization" and a wealthy white entrepreneur (Mutopo & Chiweshe, 2014, p. 129), this company has privately appropriated water from Mwenezi and Chisumbanje communities in Zimbabwe.

In the same communities, the location of a US$600 million ethanol plant by a company called Green Fuels, has blocked off communal water sources for corporate use (Chiweshe & Chabata, 2019; Mutopo & Chiweshe, 2014). Rivers such as Injelenga, Duvi, and Sosonye have all been marked as private property of wealthy commercial farmers, mostly white, who contend that by their paying a price for the water bodies, the riparian water rights lapse and, in their place, a commercial private property rights system of water is erected and policed against community access (Mutopo & Chiweshe, 2014, p. 132; Chiweshe & Chabata, 2019). So, both national authorities and private corporations have been involved in privatizing water bodies previously regarded as part of the commons.

While this "great transformation" – in Polanyian terms (Polanyi, 1944/2001) – abounds in West Africa, in countries such as Senegal, Côte d'Ivoire, Gabon, and Niger – all materially poor countries (Bayliss, 2014), the transformation in Ghana, one of Africa's most stable and respected democracies, is notable. What makes Ghana a particularly interesting case is that the state is oriented to achieving an agrarian expansion led by mechanized farming. Coupled with its relatively stable governments, Ghana has become an attractive destination for investors. In these debates about land and water grabs (Boamah, 2014a, 2014b; Kwoyiga, 2019; Obeng-Odoom & Gyampo, 2017), the Ghanaian state could, in many respects, stand with the land grabbers. The state simultaneously praises land and water grabbing for turning the lives of farmers around and for supporting the people of Ghana (Ahwoi, 2010). These positions are controversial, although the Ghanian state also provides avenues such as the courts for contesting its claims (see, for example, Obeng-Odoom & Gyampo, 2017). Generally, the Ghanaian authorities welcome engagement and debate. Their institutions are even more encouraging. So, both analytically and pragmatically, more careful analysis of outcomes is warranted.

The Lived Experiences of Water Markets

Using the criterion in quadrant (a) in Figure 7.1, the marketization of the commons, or the marketization of liquids, can be said to have had demonstrable positive impacts. In one study, the technology introduced for the private sector-run large-scale plantation in Yendi reduced run-off by 30 per cent compared to pre-technology arrangements. Indeed, in the cases where jatropha has been grown on marginal lands, there have been benefits in terms of how reduced run-off has led to reduced erosion, and some marginal and poor lands have been restored or reclaimed (Kizito et al., 2013, p. 355). Such restored land will, therefore, add to the stock of fertile land in the Yendi area, potentially helping to expand the livelihoods of the people. Boamah (2014a) has also suggested that the process of land allocation serves two symbolic and material purposes. Symbolically, the chiefs who are responsible for assigning land have used the process of land allocation as a symbolic way to assert their power of control over land. Economically, the chiefs use the process as a path to formalize land and make it more acceptable as an investment vehicle for rural development. In a different study, Boamah (2014b, p. 329) showed that, in 2010, local residents of Nsonyameye Village benefited from ScanFarm Ghana Ltd (previously called ScanFuel), a biofuel/maize/soybean company affiliated with ScanFuel AS of Norway but currently working in Ghana. According to Boamah (2014b), the company allowed residents to collect the leftover of its bumper maize harvest. In Northern Ghana, the Integrated Tamale Food Company provides training and inputs for local farmers interested in joining its outgrower scheme, for which 934 people have been employed (Tsikata & Yaro, 2014; see also Dinko et al., 2019).

None of this evidence, however, proves that investment is a natural consequence of marketization of the commons. Indeed, the evidence neither specifically demonstrates that job creation is intrinsic to water markets nor shows that the investment that created the jobs identified could only be undertaken through privatization and marketization. As Figure 7.2 shows, there can be job creation and decent livelihoods under a commons system of property rights. With the support of the state, local systems can be scaled up and substantially widened to provide even more social and ecological protection. Similarly, public ownership of resources does not debar investment, innovation, and job creation. Rather, as many articles (e.g., Obeng-Odoom, 2013c; Potts & Hartley, 2015) in *Review of Social Economy* show, the social economy as the embeddedness of the economy in society and environment (Polanyi, 1944/2001) produce much innovation, investment, growth, and change.

Indeed, the work of Schumpeterian political economists to which the *Journal of Evolutionary Economics* is devoted (see, especially volume 25, issue 1, 2015) has demonstrated concretely and in diverse settings that it is the social economy, not the market economy, that brings about the more lasting social change. Advocates of marketization have not shown empirically why private property and marketization are prerequisites for investment and why common property is an obstacle to this same investment, this particular line of thought can be abandoned as an ideological normative, especially when projects that have given little reward compared to the social, economic, ecological, and even spiritual benefits of land and water (quadrant [b] in Figure 7.1) are more substantially in evidence. For instance, jatropha companies in the Pru district in Ghana provide 120 low-income jobs (US$50/month) for 780 hectares of land leased out. While employees like their jobs for the security of income, they see that it is better as a complement rather than a substitute (Schoneveld et al., 2011). Other marketization-related employment has been reported in the literature (e.g., Williams et al., 2012), but the rewards are low and the employment mostly on a casual basis.

In terms of quadrant (c) in Figure 7.1, contrary to promises about superior accountability mechanisms, there is much evidence of unaccountable water management practices. Some companies have not declared or applied for water permits, although they intend to use the water resources on their huge parcels of land. The research of Kizito and his team (2013, p. 342) shows that for Kimminic Estates Limited (KEL) operating in Kobre, "water rights were not explicitly stated in the land deals, but KEL has started exploiting water on the leased land." Solar Harvest (SH) in Yendi took a lease of land for 50 years, and although the lease did not make explicit reference to water, "SH is going into full-scale irrigation ... SH plans to pump water from the White Volta River in the future for irrigation" (Kizito et al., 2013, p. 342). Further, apart from Kimminic Estates Ltd, the other companies have switched from the cultivation of agrofuel plants to the cultivation of other crops that are more dependent on water, without informing state bodies. In turn, so-called environmental impact assessments carried out are rendered less useful as they do not capture unreported changes in contract terms.

Those impact assessments, to be sure, were never externally verified; they were prepared by the companies' consultant firms and so present only a one-sided story of corporate capital. From the perspective of the communities, the impacts of the agro-water activities on livelihoods were neither accounted for nor reported (Williams et al., 2012). The evidence of access to water is mixed. While Nibi's (2012) study found that one agrofuel company has constructed two dams for the community to now access water at a shorter distance, Williams et al. (2012) and, more recently, Dinko et al. (2019) found a weakening of access to water for women, especially, as they have had to travel longer distances to fetch water.

These studies, however, concur and confirm that women's access to firewood for the home has reduced as nearby water sources, banks, and woodlots are enclosed and appropriated. As only 16.2 per cent of the population in Ghana has "water on premises" (Sorenson et al., 2011, p. 1524), the increase in time spent looking for water is likely to be a problem for most of the population. Also, as men are given priority over women and children whenever access to water in Ghana becomes limited (Dinko et al., 2019; Sorenson et al., 2011), it can be argued that market-state induced scarcity of water has important health implications for women and children. Similarly, this trend of weakening access to water imposed by the marketization of the commons not only unleashes uneven ramifications for women and children in society but also triggers health problems for them. Other weaker groups also bear the cost of marketization of the commons. Recent research (Dinko et al., 2019; Nibi, 2012; Williams et al., 2012) shows that migrant farmers have been particularly adversely affected by the marketization of the commons.

Quadrant (d) type of outcomes (see Figure 7.1) also require scrutiny. Women in Ghana have experienced the losses differently from men. Land regarded by men as fallow is used by women to grow some vegetables (Obeng-Odoom, 2014c), so when these fallow lands are marketized, women are disproportionately affected (Dinko et al., 2019; Schoneveld et al., 2011; Wisborg, 2012). In addition, with investors working most often through male-centric systems, including the chieftaincy institution dominated by males (Boakye & Béland, 2019; Boamah, 2014a, 2014b; Sackeyfio-Lenoch, 2014), existing gender-based inequalities have been accentuated, at least in the Dipale and Kpachaa communities in Northern Ghana, where the Integrated Tamale Food Company acquired 552 hectares of land, part of which was devoted to the cultivation of *dawadawa* trees, traditionally reserved for women's livelihoods. The land acquired became a commercial mango plantation on outgrower terms. Yet the best jobs arising from the project went to men. Indeed, of the 600 casual jobs with poor pay and less social protection generated by large-scale farming activities, women took 80 per cent; men took about 70 per cent of the 255 permanent jobs with better pay (Tsikata & Yaro, 2014). The highest pay and perks from the large-scale projects are, however, appropriated by the agents of the state, the industry of consultants, Indigenous elite, and corporate capital (Dinko et al., 2019; Williams et al., 2012; Wisborg, 2012; Yaro & Tsikata, 2013). So, not only is the menace of privilege expressed in terms of expropriating common wealth for private gain or failing in attaining its own expressed goals, but it also forces weaker majorities into grinding socio-economic poverty and further degrading conditions.

Still on the quadrant (d) analysis, unlike the egalitarianism embodied in pre-colonial commons, much state investment effort has strongly favoured transnational companies investing in agrarian activities underwritten by prospects for exclusive water rights. Transnational corporate entities have found the "plentiful" and exclusive rights to the use of natural water bodies attractive. The availability of and exclusive right to access and use the water from White Volta River in Northern Ghana, for example, were crucial factors in the decision of the Integrated Tamale Food Company to locate in the village of Dipale (Yaro & Tsikata, 2013). Other agrarian companies such as ScanFarm Ghana Limited, KEL, and SH Limited have obtained land deals with ready and exclusive access to water, for which they obtained no permit (Williams et al., 2012).

The new water right holders do not commit to conservation, let alone the expansion of the ecosystem to which they now hold exclusive rights. Indeed, invoking Tony Allan's (2003) idea of "virtual water," the idea is that when farm produce are exported in trade outside where they are grown, it is to be understood that what has been exported is not only produce but also water. This "embedded water" idea suggests that fruit producers and jatropha cultivators in Ghana who export their end products outside the communities from which they took water to grow the groups are virtually divesting the communities of their water resources. It has been suggested that "a flower is 90% water" (Grant, 2015, p. 205) in one African community, but more precise calculations of how much water is embedded in which crops at what time will have to be conducted. For now, however, it can be argued that the promise that marketizing the commons leads to more effective water management for people, economy, and society is yet to be seen in practice.

Interpreting the Outcomes of Marketizing the Commons and Deciding Whether Water Should Be Commodified

Advocates of the conventional wisdom could argue that the challenges that I have shown do not arise from water markets per se but with the implementation and nature of the property rights system in Ghana. From this perspective, if the property rights system is not clear or has its own difficulties, this will likely translate into suboptimal operation of water markets. The focus is, thus, on the structure and enforcement of the property rights system and not the market itself. If the private interests and companies do inform state bodies about their actions (such as switching crops and reporting how much water is going to be taken) and the state is able to compel them to switch back to growing the contracted crops, all the problems discussed will disappear. Or, if not, there will be some legal processes to address the issue of non-compliance. This view is historically located in the idea that social problems should all be pinned to institutions other than the market and a "market fail-ure" can only be because of minor implementation problems not fun-damental issues; hence, more or better marketization is what is needed (see, for example, Anderson & Libecap, 2014; Arezki et al., 2015).

In theory, it is possible to use tighter regulation to improve quad-rant (a) to (c) type outcomes, at least, but there is a Catch-22: the more regulation, the less attractive a destination is for capitalist expansion. Indeed, even World Bank economists Arezki and colleagues (2015) and others (e.g., Grant, 2015, pp. 240–241; Obeng-Odoom, 2020) have shown that one major reason white and wealthy "land grabbers" are fleeing South Africa to take land elsewhere in Africa is increasing black workers' rights and fear of tighter regulation of agro-based invest-ments. This interpretation misses the *raison d'être of* capitalist strategy: land is "grabbed" as a business plan to attain exclusive possession of water and this is actually considered to be a "conducive business envi-ronment" (Pearce, 2012, pp. 102–103). But even more fundamentally, more regulation without a structural change in water governance will imply ignoring quadrant (d) concerns, which can only be addressed by considering water as a human right.

"Water," the Committee on Economic, Social and Cultural Rights has noted in General Comment 15 (1), "is a limited natural resource and a public good fundamental for life and health." According to the committee, "the human right to water is indispensable for leading a life in human dignity." For the committee, water "is a prerequisite for the realization of other human rights." The right to water is one of the most fundamental of all human rights. The now expired Millennium Development Goals, particularly Goal 7 (target 7c), prioritized access to safe water as the main indicator of progress in water management. The SDGs essentially maintain this focus in Goal 6, albeit with some cosmetic changes in wording, while emphasizing "acccess." Yet access is not enough: water is a right, one to be had as an entitlement. Accord-ing to the committee, "the human right to water entitles everyone to sufficient, safe, acceptable, physically accessible and affordable water for personal and domestic uses. An adequate amount of safe water is necessary to prevent death from dehydration, to reduce the risk of water-related disease and to provide for consumption, cooking, per-sonal and domestic hygienic requirements" (General Comment 15 (2)). The growing awareness at the highest levels of international develop-ment that water is, indeed, a human right, peaked on 26 July 2010, lead-ing to a UN Resolution (64/292), which is a full endorsement of the committee's famous comment 15. This right to water, which is linked to the right to good sanitation and food security, has been upheld and further endorsed by the UN Human Rights Council in Resolution A/HRC/15/L.14 (Tignino, 2014, pp. 383–402). The UN position tends to be state-centric, however, and often disconnected from broader Indigenous rights to the commons.

The pre-colonial arrangement in Ghana, which has much potential for holding chiefs to account, while upholding community values, ethics, and ethos, can be a better basis for an alternative and comprehensive water rights system (see Figure 7.2). Of course, the issue of limited information for some chiefs and other traditional leaders was a major challenge in pre-colonial Ghana. However, the present system of vibrant media activities, the existence of modern surveying and measurement, and a new class of chiefs, some of whom hold postgraduate university degrees, can all help to curb the problem of limited information on measurement with which pre-colonial chiefs contended (Austin, 2005). Patriarchal structures, worsened by colonial practices such as the penchant for British colonial officers to consult mostly males for advice on "custom" (Walker, 2002), codification, and neoliberal adjustment programs, remain, although the potential to loosen their grip exists in the form of greater media and civil society pressure together with legal reforms (Amanor-Wilks, 2009).

A few chiefs such as the *Okyehene* have set up environmental foundations that adapt Indigenous methods for "modern" ecosystem management in their chiefdoms. Such Indigenous "counter-revolutions" have been promising enough to draw the support of global environmental civil society groups (Eshun, 2011). A similar project is ongoing in Berekum, where traditional authorities led by the paramount chief of that traditional area have generated much international interest in their use of Indigenous knowledge for water and environmental management (Awuah-Nyamekye, 2013). So, there is hope for a fundamental alternative to the marketization of the commons, especially water – a kind of counter movement or "double movement" in the Polanyian sense (Polanyi, 1944/2001).

The question, of course, is how to build on these alternatives to reverse the sweeping market changes. The goal is not to return to pre-colonial forms *in toto*. Rather, it is more useful to strengthen Indigenous institutions for new and inclusive governance of the water commons. Lydia Kwoyiga (2019) documents the persistence of such a system in farming communities in Northeast Ghana. Reviving the *asafo* groups, immediately recognizing customary rights to water, and regarding access to and control of water as a human right can also form part of an integrated strategy. Although dormant, the *asafo* groups, for instance, are not extinct (Gyampo & Obeng-Odoom, 2013; Paller, 2019). Besides, there are contemporary signs of grassroots progressive forces. For instance, in Kpachaa community, some dispossessed farmers express their discontent by burning the crops of investors (Nibi, 2012). In Dipale, also in Northern Ghana, some natives have resisted by abstaining from sharing local knowledge about how to prevent fires or not helping to quench them when they start in a region that is prone to such fires (Pickbourn, 2020; Yaro & Tsikata, 2013).

In southern Ghana, too, there have been murmurings about the top-down nature of land and water deals. These concerns culminated in a demonstration by some natives of Agogo in 2010. Another demonstration was staged in 2011 to protest the dispossession of land (Wisborg, 2012). In terms of using state institutions for progressive ends, the registrar of Knutsford University College in Ghana has filed a suit at the Supreme Court seeking the following reliefs: that the Government of Ghana be forced to live up to its duty of holding land for the public good; that the Government of Ghana desists from giving land to TNCs beyond lease-hold terms of 50 years; and that steps are immediately taken to return land to the control of customary leaders (Issah, 2013; *Nana Oppong vs. Attorney General and Minister of Justice*). Judgment was yet to be given at the time of writing in this case, but international evidence (e.g., Boyd, 2012; Jeffords, 2015; Jeffords & Minkler, 2016; Manning, 2018) suggests that a positive constitutional support for communal rights could provide greater protection for the commons. At this stage, only time will tell whether these protests – grassroots, intellectual, and legal – can be coordinated and sustained to bring about decisive change in the direction of small-scale, community-based hydro-agro development.

Conclusion

The recent surge in the marketization of the commons, especially the creation of water markets and the commodification of water in Africa in response to Conventional Wisdom, raises three questions. First, were there markets in the beginning? If so, how have they transformed and if not, how did markets arise and evolve over the years? Second, what are the outcomes of such markets for people, their livelihoods, and their environment? Third, how should we interpret the outcomes of water markets and decide whether water should be commodified at all?

Existing challenge by concerned analysts as part of the Western Left Consensus has highlighted the humanist case against such marketization. The more Radical Alternative developed in this chapter has tried to provide a more holistic analysis that uses multiple sources of data to probe absolute, relative, and differential/congruent outcomes, as well as the opportunity cost of the current water property rights regime.

On these bases, the chapter finds that, although the Conventional Wisdom is that markets have arisen because of the inferior nature of Indigenous or customary systems and that water markets offer great economic and ecological fortunes for African countries committed to the mimicry of formal, individualized water rights systems, the empirical evidence is more intricate, showing the systematic and directed efforts to usher in and expand markets.

While clearly some job creation and even local food distribution are associated with some transnational hydro-agro projects, overall, there is no necessary congruence between the promises made by new institutional economists and how communities experience water markets. Indeed, the private model of property rights has worsened the distribution of water resources, not only within different property relations in Africa but also between diverse property relations in Africa and across the world. Water markets have been responsible for much displacement and created much trouble for communities and for nature. Tighter regulations for the use of inland and transboundary water sources might temporarily halt the displacement of communities sparked by the capture of community water resources, but only one fundamental change can guarantee community well-being. As this chapter has shown, this change is to regard the access to and community control of water as a human right and as *res communis*.

This alternative system worked to support the economic, social, ecological, and even spiritual well-being of pre-colonial Ghanaians. Indeed, the lessons are of a general applicability, as the evidence in Lars Sundström's (1974) seminal – but poorly known[2] – study, published as *The Exchange Economy of Pre-Colonial Tropical Africa*, shows. Specifically, that seminal study covers West, Central, and East Africa. Sundström investigates how the economy works, focusing on institutions such as gifts, taxation, trade, credit, hoarding, and money. His systematic investigations seek to concretely detail how the pre-colonial economy worked before colonization, so the eighteen and the nineteenth centuries are of primary interest.

A key finding of Sundström's study is that the economy was based on gifts. In this sharing economy, gifts were abundant. However, how much people gifted was based on their ability. Gifts played the role of ensuring social inclusion, not economic accumulation. If this institution worked to ensure harmony and more egalitarian societies in Africa (see

---

2 Such a path-breaking study has been cited only 90 times since 1974, according to Google Scholar, https://scholar.google.com.au/scholar?hl=en&as_sdt=0%2C5&q= The+Exchange+Economy+of+Pre-Colonial+Tripical+Africa&btnG= (accessed 9 December 2018).

Ojong, 2020; Obeng-Odoom, 2021) and black societies elsewhere in the world (Gordon Nembhard, 2014a, 2014b; Hossein, 2016, 2018), it was the institution of trade that brought even more intriguing outcomes

Contrary to the view that pre-colonial Africa was a "Dark Continent" isolated from others and populated by inward-looking "tribes," Sundström's study (see Sundström, 1974, pp. 13–20; pp. 45–64) shows that both internal and international trade were widespread. Specific institutions about trade made it personal rather than impersonal. Social regulations shaped how trade was done. For instance, the principle of *caveat emptor* which shifts the burden of ensuring quality goods are exchanged to consumers was of limited application (see, Sundström, 1974, p. 21). The historical evidence shows that Europeans found such practices too time-consuming (see, Sundström, 1974, p. 22), although it provided a strong basis for an inclusionary society.

Consider the institutions of hoarding and taxation. Hoarding was prohibited. What hoarding existed was only meant for burials. So, accumulated wealth was buried. Sundström's evidence (see 1974, pp. 116–121) shows that the motive for burying rather than using accumulated wealth for economic advantage was partly to ensure that future generations did not obtain unfair advantage, partly to help the hardworking individual in the next world, and partly to avoid any diseases that could be transmitted by sharing the wealth of the departed. Taxation had a more directly redistribution function, as it fell on surplus and chiefs actively sought to spend the revenues from tax on providing public goods and meeting the needs of weaker people in society. Indeed, as Sundström (1974, pp. 6–7) shows, unlike elsewhere, taxation in Africa was not to force the population to help in the spread of money.

Money itself was not needed in the economy (on this issue, see also chapter 2, especially "Money, Debt, and the Origins of Private Property"). Sundström's evidence shows that bartering was common. Yet in contrast to the claim that barter was such an inconvenient system, Sundström shows that it worked "smoothly," "profitably," and "easily" (see Sundström, 1974, p. 68). Europeans tended to struggle in the economy, especially if they had no patience. For those Europeans who endured and adapted, barter brought them considerable opportunities and, hence, they opposed the introduction of money (see Sundström, 1974, p. 69). Money was introduced, nevertheless, but mainly to help Europeans, not the Africans, according to Sundström (1974, pp. 106–107).

Barter was not an underdeveloped form of exchange, and it did not pose insurmountable challenges. Barter-related problems of finding suitable goods to exchange for owned goods were addressed using institutions of credit and deferred payments, both of which had distinctive features. Credit, for example, was not usurious (see, Sundström, 1974, p. 39; compare colonial and imperial credit discussed in chapter 2 of this book). Both credit and deferred payments were enforced by moral, rather than legal, sanctions. In this sense, even after the introduction of money, its utilization was closely linked to its use value (see Sundström, 1974, pp. 111–116).

This commons-based Indigenous system in Ghana and elsewhere in Africa also applied to other types of land, not just water. Sundström (1974, see chapters 4–7 of Sundström's book) documents the principles in the salt, iron, copper, and brass economy. That this commons-based system led to prosperity in advanced civilizations is also well known, as the seminal study by Cheikh Anta Diop has now established in the case of pre-colonial Egypt (for a detailed discussion, see Diop, 1967). Across Africa, and elsewhere in the world, considerable evidence (see, for example, Gordon Nembhard, 2014a, 2014b; Hossein, 2016, 2018; Ojong, 2020; Boonjubun et al., 2021; Obeng-Odoom, 2021) shows that commons-based economic models guarantee inclusive and environmentally enhancing prosperity. What, then, are the bigger lessons to learn from this study for the future of the commons?

PART D

The Future of the Commons

Chapter Eight

Concluding Remarks: Towards a New Ecological Political Economy

The volume and the rate of publications on the commons can be confounding. Yet most of them can be grouped into two categories: The Conventional Wisdom and the Western Left Consensus. This book has tried to make three contributions that constitute the Radical Alternative. First, it has empirically demonstrated the myth of privatizing nature. Using markets to resolve social problems – as the Conventional Wisdom advocates – is ineffective. Markets tend to complicate social problems. They also create new social tensions. They generate, maintain, or extend the contours of socioecological inequalities through their creation, maintenance, or extensions. Often these three forms of marketization are intertwined in producing these socioecological differences. Consider, evictions. They arise when private property is institutionalized and enforced, a process that, as we saw in chapter 4, reflects wider slave and neocolonial processes.

Second, the book has provided a systematic defence of inclusive prosperity from common property. As maintained throughout the book, this common property is better considered as "land," not the Western Left Consensus idea of omni commons. When common property is maintained through commoning, progress comes without poverty. Commoning here means something rather different from what Peter Linebaugh (2008) means. Instead, commoning land, including using land value tax, maintaining non-privatized land as common land, and preventing the intergenerational transfer of monopolized land, can generate many of the conditions that will free labour from the bondage of taxation and exploitation. Commoning land also enables labour to flourish because it restores to labour its rights to land, which can be put to liberating socioeconomic uses. Similarly, commoning land provides the surest protection for nature without undermining the livelihoods of labour and socioecologically sensitive production that is neither colonial nor colonizing.

Third, as this chapter draws out, this book has made the case for developing "rent theft" and "just land" as concepts, and "the Global South" as a concrete research approach in ecological political economy to decolonize the methodologies of the Conventional Wisdom and the existing Western Left Consensus.

Much of the attempt to reclaim the commons has centred on politics and political activism, without seeking to reclaim more compelling analytical political economy of the commons. Although action is important, activism informed by problematic analysis is likely to be wrong-headed. It is this neglect that led K. W. Kapp (1971) to the following conclusion:

> The success of any program of environmental control depends ultimately on a correct analysis of the manner in which social costs are incurred, on the adequate assignment of responsibilities for them, as well as on the effectiveness of the practical and institutional measures adopted to overcome them, and finally on the adequacy of the funds appropriated. A superficial and, hence, incorrect analysis is likely to lead to ineffective measures of control, and even a correct program will see its chances of success jeopardized by the allocation of inadequate funds and the lack of appropriate institutional arrangements. (p. ix)

This book has tried to correct this analytical gap in the literature on the commons. Its contributions are both analytical and empirical and could potentially inform both policy and political action. In this three-part last chapter, the key challenges in the debates on the commons are summarized under "Requiem for Conventional Wisdom and the Western Left Consensus." Next, the possibilities for policy action are highlighted under "Prospects." Finally, under "Towards a Just Ecological Political Economy," analytical lessons from the study are developed for future research.

Requiem for Conventional Wisdom and the Western Left Consensus

The commons debates are currently stuck in a dialogue between the Conventional Wisdom and the Western Left Consensus. Of course, some (e.g., Zhang, 2017) position themselves between these schools of thought. What this book has done, however, is to develop a Radical Alternative that goes beyond these binaries. As demonstrated in chapters 1, 2, and 3, the theories of the commons have made no real progress beyond these polarized confinements. Hardin's articulation gave the theory greater visibility, perhaps, and Ostrom's intervention was widely and loudly praised, despite making no real departure from Hardin. Indeed, by 1998, Hardin and Ostrom were indistinguishable. In turn, the attempts by the Western Left Consensus to simultaneously praise and seek to expand the work of Ostrom put their alternative in an awkward position. While parading as "radical," the steps this Western Left Consensus proposes are incapable of leading to its own vision. Top-down, static, and dualistic, this vision is inadequate.

Indeed, there is an insidious consensus between the mainstream and Western progressives. This alliance is hiding in plain sight; anyone looking closely at the research on the commons can easily see it. Consider the issue of land. As this book has shown, the Conventional Wisdom holds that it is a factor of production much like anything else. The Western Left Consensus agrees. In its conception of the commons, land must be commoned much like other factors of production, so the unique properties of land – such as rent and how land has been shaped by slavish processes and colonially instituted and maintained neocolonially – are poorly understood by the Western Left Consensus. Its commitment to land is both half-hearted and one-sided. Indeed, its interest in "sustainability" and "degrowth" is bankrupt because, as shown in chapter 7, without recognizing what the Africans call "just sustainabilities" and "just transition," the Western Left Consensus shows its insensitivity to global injustices and total commitment to destroying the true social costs of private enterprise.

Growth may be at the roots of social costs, but which growth? What about inequality and social stratification? Neoliberal private enterprise, adherents normally contend, but how specifically does growth lead to evictions, to waste, to congestion, to pollution, to biodiversity loss, and to world-scale ecocide? The Western Left Consensus has no clear answers other than to claim that growth for profit and not for human need is the enemy, but this does not answer the practical questions raised about contradictions outside the factory. Neither is an appeal to butterfly effects very persuasive. Therefore, the demand for the end of unsustainable growth only is a leap of faith.

The Radical Alternative, developed in this book, is centred on the analysis of land. As shown in chapters 1 to 3, land and landed relations provided the context for the historical origin of money. Today, land provides the foundations for the power of TNCs while maintaining the power to lay the foundations of economic growth (see, especially chapter 6). Urban land is at the heart of the idea that cities are growth machines (see chapter 4). Basically, at the advanced stage, growth can be used to explain everything. However, claims that growth helps the environment through technological advancement are problematic. As shown in chapters 4 and 5, technological fixes could exacerbate, rather than help, the crises of the environment. The framework of land has enabled us to see the limits of technological advancement in a new light. Public transport could be developed, green cars created, and multimodal transport and urban consolidation encouraged, but looking from the land, we can see how "advances" work to the benefit of landlords. This form of landlordism unleashes socioecological problems such as evictions, widespread inequality, pollution, and biodiversity loss.

This book has shown that monopolizing land allows a small elite to control the world. This 1 per cent also benefit from the social problems unleashed through the privatization of nature. Standing this logic on its head by commoning land can reverse many of the social problems. It will also dismantle the structure that makes the present social problems possible (see chapter 5). The positive case is that commoning land can allow for prosperity for the present and for posterity. New business will flourish in ways that are not monopolistic. In this environment, lazy rentierism will no longer be rewarded.

As chapter 7 shows, pre-colonial Africa gives lucid examples of how such societies can work. Evictions are rare, as land itself is shared. Strangers are welcome to access land. They can share in the fruits of the land with the land-giving community. Underlying institutions such as *abunu, abusa,* and *Karaafo* have not previously been applied outside of agrarian contexts but, following arguments made in this book, they could. As it was in those days when the state used rents for social purposes, so it can be in our time, too: many of the institutions were socially created, not biologically determined and programmed only for a particular epoch. Indeed, as the book has shown, many of these institutions constitute to exist and to flourish. Rents can be put to social purposes. Enhancing open access to public parks, ensuring public technology, and guaranteeing public education and health as part of the wider social interventions discussed in chapter 6 are both possible and desirable.

The contention about corruption, often made by advocates of the Conventional Wisdom, is one of the weakest and lamest arguments about the state in Africa and the Global South more generally. The features of state reflect the social system of which they are a part. Just as capitalist states take on capitalist features, and capitalist states can drive the march towards capitalism, so it is that in a commons society, the state takes on commons features through a process of commoning and can drive the march towards the commons – as borne out by historical experiences (see also Schläppi, 2016, 2019) and the detailed exposition in various chapters, but especially in 6 and 7. There is no need for repetition here, except to emphasize that there are signs of African autonomous action.

Prospects

The prospects for transforming the current system appear quite promising. Discouraging speculation is one way. Reducing monopoly is another. Urban land values can be taxed to achieve these ends, to miti-gate stratification, and to enhance public revenue. For "strangers" or foreigners, in addition to paying these taxes, the rents from production could be shared using local institutions (*abunu* and *abusa*) of rent shar-ing. Rents – in the form of oil – could be shared 50–50 (*abunu*) or in the ratio 1:3 (*abusa*) depending on how much investment is made by the strangers (e.g., TNCs). This Indigenous approach applies regardless of whether there is a bumper harvest. In this way, strangers could be included in the commons, but they, in turn, would have to support the commons.

As these steps begin to bear fruit, income taxes could be reduced. Labour could then be freed and encouraged. Along with public sup-port, labour-based initiatives could be expected to blossom. The result-ing possible increases in earnings could stimulate more sustainable local production and distribution. This cumulative cycle of virtue con-stitutes a rather different approach, quite distinct from the Western Left Consensus, which is centrally focused on trade unions and unionizing without engaging with debates on rents and how such rents undermine labour (for reviews of the Western Left Consensus on the liberation of labour, see Stevis et al., 2018). The Radical Alternative, however, liber-ates labour from exploitation. Apart from the additional incentive to relax, to innovate, and to be autonomous, new social states could also emerge. Survey evidence (Friedrich-Ebert Foundation, 2011; Obeng-Odoom, 2015a, 2015b; Wilde et al., 2013) suggests that this alternative is both popular and preferable.

These developments are not only local or localized. Instead, they are being increasingly regionalized. As I explain elsewhere (Obeng-Odoom, 2020), in 2010, the Kolongo Appeal was made by peasant groups in Mali who tried to provide organized local resistance to land grabs. Subsequently, the Dakar Appeal led to a global conference of peasants in Mali in 2011. The resulting "commitment to resist land-grabbing by all means possible, to support all those who fight land grabs, and to put pressure on national governments and international institutions to fulfil their obligations to ensure and uphold the rights of peoples" (Nyeleni Conference, 2011, p. 2) provides much backing for the Radical Alternative.

As argued throughout the book, central to the growth imperative is the pursuit of rent. This tendency is what drives the incessant creation of credit and money (chapters 1 and 2), absentee ownership, specula-tion, land grab, territorial expansion, and monopoly. The structural "growthmania" in capitalism, which plays out in distinctive ways in cities, often regarded as "growth machines" (Molotch, 1976), controlled by landed and capitalist monopolists (chapters 4 and 5) is in lockstep with the private property system. This catechism tries to monopolize the very earth on which we live and its natural gifts – including water – for common people (chapters 6 and 7). In the process, structural growthism creates widespread evictions, spatial segregation, and other inequalities; it pillages the environment while it tries to rapaciously dig up all its resources and, in the process as well as in outcomes, destroys social bonds. Thus, a private-land-based economic system could create growth but in ways that are fundamentally inconsistent with its own vision of the good society. It is, thus, a myth to privatize nature.

To switch the system to a steady state economy, rentier capitalism with all its institutions must be replaced. A commons-based system – in which land is a commons – provides a rather different potential. It could trigger new structures and processes to transform society. Indeed, this new economic system is rather distinct from existing Western Left Consensus alternatives, such as reformist capitalist systems, with new demands on individuals to eschew consumerism. My proposed Radical Alternative differs from market socialism, and state socialism. These alternatives cannot transform wage labour, the nature of money, the contents of markets, TNC-based profits, or the nature of the state – key pillars that undergird the relentless drive for growth (Exner, 2014). More fundamentally, the existing strategies by the Western Left Consensus ignore the element of rent, private property in land as its core vehicle, and bonded labour as one core consequence (Obeng-Odoom, 2018). The commons-based system, defended in this book, provides the stepping stones not only to avoid unsustainable growth but also to reach a just and stable economy. This alternative also seeks to resolve social and economic inequalities through a new social state and institutionalizes mechanisms to prevent their recurrence, such as social spending and common property (see chapters 3, 6, and 7, for example).

This commons-based mode of production and organization is distinct from the focus in both the Conventional Wisdom and the Western Left Consensus. It engages these notions of commoning, of course, but it is more successful in demonstrating the myth of privatizing nature and more coherent in defending ecologically sensitive and socio-economically inclusive prosperity through common property for posterity.

Like K. W. Kapp (1971), I consider this exercise in concluding my arguments as a point of departure rather than arrival. So, additional policy and empirical details will have to be worked out by other students of political economy working on topics such as general debates about "the gift," specific debates on the gift of land, and how these two themes can shed light on the wider implications for the commons. For such students, it is important to further clarify the concepts and tools that could enhance their political economy.

Towards a Just Ecological Political Economy

Existing ecological political economy (for a detailed critical review of its features, see Rosewarne, 2002) has provided a strong challenge to environmental economics. It also provides a source of inspiration for complementary analytical paradigms. Within heterodox economics and political economy more widely, ecological political economy has kept alternative economics relevant by working the environment into a wider analysis of capitalism. In turn, it has successfully expanded the vision for political economy and become interdisciplinary by reaching out, for example, to feminist economics.

However, ecological economics has tended to be trapped in an "environment-ecology" speak, without embracing a wider analysis of social problems. K. W. Kapp (1971) was prescient when he said,

> One word of caution: The increasing use of the terms "environment" and "ecology" in recent discussions of social costs is to be welcomed, provided these terms are interpreted in a sense sufficiently broad to include not only the impairment of the physical environment but the impairment beyond certain definable threshold levels of the aggregate of all external conditions and influences affecting the life and development of human beings, human behaviour, and, hence, society. Only in this way will it be possible to counteract the widespread but false impression that we are confronted only with a problem of ecology in the narrower physical sense of the word. (p. x).

Today, only a few writers in the ecological political economy tradition engage the writings of scholars such as Julian Agyeman who advocates "just sustainabilities." Engaging the work of Southern thinkers such as Wangari Maathai (see chapter 6) is much less frequent than it could be. Yet many of these analysts do not have good, first-hand experience of the social problems in the Global South. Focused almost exclusively on providing a critique of capitalism without considering its diversity in multiple locations and how different races, genders, and other identities experience it (Crenshaw, 1991; Gibson-Graham, 2006a, 2006b; Hill, 1961, 1966; hooks, 1982; Showers, 2014), these (ecological) political economists remain well intended but quite narrow. Many are advocates of the Western Left Consensus. There are exceptions. Daniel Bromley, for example, who has been a long-standing advocate of looking at property, land, environment, and development (see, for example, Bromley, 1991, 1992, 2019). Generally, however, ecological political economy has paradoxically remained weak on the political economy of land. That is a serious neglect for, as shown in this book, commoning land has the potential to become the foundation of global-social, economic, and ecological prosperity.

So, if ecological political economy is going to be more relevant, then it must not be just the ecological political economy that it is now. It must be a just ecological political economy. The journey from here to there must involve taking concrete steps, such as embracing the ideas of just sustainabilities and just transition more seriously. This new eco-logical political economy should also knit together a new approach to political economy centred on the three interdependent concepts: rent theft, just land, and the Global South.

Rent Theft

As a political-economic concept, rent theft connotes the private extraction and appropriation of rent created by society at large. It is a theft because it is an appropriation of something that does not belong to the appropriator. Even if the appropriator makes some contribution to it, by appropriating all of it, it becomes a theft. It is a social wrong not only because it is an ethical wrong but also because it creates concrete social problems and socioecological crises. It is a brake to production and a prelude to recurrent crises and continuing global war, often mistakenly called "civil" wars.

Unlike "wage theft," which is localized to labour-capital relations (for a detailed discussion, see Doussard & Gamal, 2016), rent theft affects all the factors of production and can be the precursor to wage theft. Indeed, when the gains of labour are swallowed by increasing rent, a rent theft has occurred. When the advancement of technology creates rents, which are then privately appropriated, a rent theft has occurred. When, via oil exploration, land rents increase and are privately appropriated, a rent theft occurs. Capitalists, too, risk losing profits to rent theft, but they seek to recover these losses from the exploitation of workers but the latter cannot shift the burden of rent theft.

Although based on Henry George's theory, neither he (see, for example, George, 1879/2006, 1883/1966, 1898/1992, 1891, 1892/1981) nor other rent theorists (e.g., Haila, 2000, 2016) have developed this concept systematically. As a political-economic idea, it is not a formula, but it would direct the attention of the investigator to rents, how they arise, in what ways they shape social, economic, and social realities, and how, by commoning rent, these realities might change. As with all political-economic concepts, the concept of rent theft might be able to animate struggles for the commoning of land not as a thing but as a social relation. In turn, activists seeking the commoning of rent could usefully pursue shares in mines – for example, land value tax and resource tax based on windfalls, alongside demanding the removal of crippling taxes on labour – rather than old-school *physical* land redistribution, which says nothing about differential rents for the same plots of land. Another decolonized prospect of this strategy is that in challenging the Conventional Wisdom centred on mining for development, it transcends the alternative revolutionary Western Left Consensus by putting the focus on non-capitalist evolutionary processes of piping down.

It follows that activists could also demand accountabilities based on rents, raising questions about the social cost of rent generation, how much rents have been extracted, and to what uses they have been put, and probing whether they address the social costs, build the potential of the society, and prepare them for a just transition.

The point of the political ramifications of this concept is not to encourage land monopolization only to extract rent later. Rather, it is to emphasize the enterprise based on real work, freeing labour from being penalized for working and, through the public use of rent for social purposes, guarantee improved social conditions not only for workers but for all. These could only be possible if land is not just land, but *just* land.

Just Land

As a concept, "just sustainabilities" has taken the debate on sustainability very far. It makes the sound and compelling argument that there is no one path to sustainability and that blacks, for example, deserve the dignity of developing their own paths to sustainability that recognize their own histories and struggles (Agyeman, 2008, 2013). The call for just transition, in the African – and wider Global South - context, is similarly spirited. These ideas can be developed further and more specifically.

"Just land" is one way of doing so. Based on the centrality of land to every aspect of African life, the concept of just land seeks to translate what is an ordinary idea into a concept for investigation. It is not just any land but *just* land that can animate the idea of rent theft. *Just land* means the return of land to the commons, the non-commodification of land in the commons, and the active nourishing of such land. This concept builds on Wangari Maathai's (2004, 2011) commitment to not just honouring what land exists but also developing the land by building trees and nurturing animals, biodiversity, and water. Just land mandates the full use of African institutions and the best institutions from elsewhere to live indivisibly in society, environment, and economy. Maathai (2004, 2011) recognized how land is destroyed by the international system, of course, but *just land* tries to use land to also rebuild that international system.

Doing so requires thinking of land as a research methodology. That could entail investigating how the international financial system, for example, generates pressures for land which then generate debt and structural global inequalities. This research approach would emphasize the place of land in creating vulnerable labour beyond the scope of unions, a central focus of the Western Left Consensus. Instead, *just land* could look to articulating the landed roots of modern enslavement on the streets of cities, in peri-urban areas, and on dump sites, a focus that could open the gates for investigating new institutions for social reform beyond unions. Politically, then, embracing the approach of *just land* could lead to liberation from wider sources of oppression and exploitation. Unionizing per se could continue, but the logical emphasis of this approach to thinking and political liberation is the commoning of land, the use of rents to provide social support, and the untaxing of labour to further incentivize them to do work that they enjoy not only in Africa but also generally in the Global South.

Global South

As a replacement for the term *Third World*, the *Global South* has become increasingly vague. It no longer accurately describes shared material conditions, as many of the former very poor countries are now quite wealthy. Although not imperial yet, many of such countries have developed institutions that could make them so. In turn, strictly speaking, it is questionable whether the Global South is still meaningful (Gills, 2016). Nevertheless, developing a collective alternative to neoliberal globalization, a cardinal reason for introducing the notion of the Global South (Dirlik, 2007), has remained an unfulfilled aspiration. The emphasis continues to be on south-south cooperation to resolve "issues of the Global South" (Dirlik, 2007). But how could the same research methodology develop an alternative paradigm?

Realizing the impossibility of arriving at a different vision using the methodology of Conventional Wisdom and the Western Left Consensus, the Indigenous scholar Linda Tuhiwai Smith has called for "decolonizing methodologies" (Smith, 2012). Progress made in this direction includes attempting to look for "Southern Theory" (Connell, 2007), thinking about cities in the Global South in their own terms or as "Ordinary Cities" (Robinson, 2006) and, actively seeking to acknowledge or revalue scholarship in the Global South under banners such as "postcolonial studies."

To consider the Global South as a methodology, it is important to engage but also transcend the existing effort to revamp the idea. Crucially, as a methodology, the Global South must shed its tendencies to be a dualistic concept that maintains an unhelpful North-South divide. This dualism can sometimes lead to important insights, but it can also be limiting. For example, pointing out that measures of well-being were developed in the West is necessary but not sufficient to dismiss them (see, for example, Mahali et al., 2018).

Instead, the Global South can become a dialectical methodology based on engagement rather than retreat. As a dialectical methodology, the Global South could weave a distinctive Southern idea such as land, which is a central identity in the Global South, into a critique of an existing thesis (e.g., Conventional Wisdom) and an antithesis (e.g., Western Left Consensus), while developing a synthesis (e.g., Radical Alternative). In this sense, Global South connotations percolate through all aspects of the methodology, not just in proposing "Southern solutions," as much of the Global South studies tend to do (Dirlik, 2007; Bob-Milliar, 2020). Thus, to say that a Global South approach should be dialectical instead of being reclusive is to say that the insights of the approach are not simply limited to the South.

As a dialectical methodology, the Global South should also be pluralist. Consider the idea of land. Even if conceptualized as a Southern construct, it has drawn strongly on Georgist ideas. That idea has also benefited from insights by an eclectic range of scholars, such as Herman Daly whose work is not so widely used in "postcolonial studies." Simultaneously, this study has also drawn on the work of leading Southern thinkers, such as Wangari Maathai and Julian Agyeman, whose stimulating insights have not percolated through the separatist and culture-heavy postcolonial studies as much as they should.

So, to consider the Global South as a methodology does not mean brushing aside all that we know, reinventing the wheel, and taking a hasty flight into so-called Southern cultures to look for insights. Rather, it is to rigorously engage and ultimately seek to transcend existing insights.

By calling that the Global South a dialectical approach, I appeal to the historical tradition in political economy. The emphasis here is not simply to be historical in chronicling happenings but also to develop alternative historiography. Currently, the "historical approach to political economy" has tended to be a substitute for the historical approach to Marxism. It is quite narrowly focused on developing the relative strength of critical Marxism over official Marxism and how, as an approach, Marxist political economy is richer, historiographically than neoclassical economics (Sherman, 1993).

In practice, even the Marxist historical approach is limited in its emphasis on class, often to the neglect of race and capital-labour struggles and the neglect of landlord-labour-capitalist struggles, even though it takes primitive accumulation seriously (Harvey, 2003). Of course, institutional economics provides, perhaps, even a more vibrant tradition. The trouble is that old institutionalism continues to struggle with "scientific racism" (Zouache, 2017a, 2017b, 2020) and its approach to race has been called "mystical" (Cox, 1945). Land and land rent are constantly racialized, so even though Georgist political economy is by far the most powerful in its dealing with the land question, neither George (1879/2006, 1883/1966, 1898/1992, 1891, 1892/1981) nor his many students (e.g., Haila, 2000, 2016) have dealt seriously with this issue. Confronting race by trying to get away from it is not a satisfactory approach, so the insights of postcolonial historical approach ought to be taken seriously. Yet even Frederick Cooper's (2014) book *Africa in the World*, one of the best in this subfield, neglects systematic engagement with capitalism and forces of neoliberalism and racism. Although neoliberalism can be quite vague for analytical purposes (Dunn, 2017), its ramifications are serious, as this book has shown, so it is difficult to justify neglecting neoliberalism and its racialized consequences. As a dialectical and historical methodology, the Global South should avoid these problems. In the positive sense, this Global South methodology should correct them while reaching out to the new field of stratification economics – developed by black and Southern thinkers (see, for example, Darity, 2009; Darity & Hamilton, 2015; Obeng-Odoom, 2020). In this sense, *The Commons in an Age of Uncertainty* develops not just another ecological political economy, but a just ecological political economy to decolonize nature, economy, and society.

References

Aadland, D. M., & Caplan, A. J. (1999). Household valuation of curbside recycling. *Journal of Environmental Planning and Management, 42*(6), 781–799.

Aadland, D. M., & Caplan, A. J. (2006). Curbside recycling: Waste resource or waste of resources? *Journal of Policy Analysis and Management, 25*(4), 855–874.

Abakah, M. (2006). *Celebration of life: Mr. Justice Moses Abakah, retired supervising high court judge, a.k.a. Kofi Baah-Kofi Boye* (pp. 4–8). Max Creation.

Acemoglu, D., & Robinson, J. A. (2013). *Why nations fail: Origins of power, poverty and prosperity.* Crown Publishers; Random House.

Acemoglu, D., & Verdier, T. (1998). Property rights, corruption and the allocation of talent: A general equilibrium approach. *The Economic Journal, 108*(450), 1381–1403.

Acey, C. S., & Culhane, T. H. (2013). Green jobs, livelihoods and the post-carbon economy in African cities. *Local Environment, 18*(9), 1046–1065. https://doi.org/10.1080/13549839.2012.752801

Adams, E. A., Stoler, J., & Adams, Y. (2020). Water insecurity and urban poverty in the global south: implications for health and human biology. *American Journal of Human Biology, 32*(1), e23368. https://doi.org/10.1002/ajhb.23368

Adams, W. S., & Mulligan, M. (Eds.). (2003). *Decolonizing nature.* Earthscan.

Adamu, Z. Y. (2012). Institutional analysis of condominium management system in Amhara region: The case of Bahir Dar City. *African Review of Economics and Finance, 3*(2), 13–48.

Adésínà, J. O. (2012). Social policy in a mineral-rich economy: The case of Nigeria. In K. Hujo (Ed.), *Mineral rents and the financing of social policy: Opportunities and challenges* (pp. 285–317). Palgrave.

Africa Progress Panel. (2013). *African progress report 2013: Equity in extractives: Stewarding Africa's natural resources for all.*

Africa Progress Panel. (2015). *Power, people, planet.*

African Development Bank, Development Centre of the Organisation for Economic Co-Operation and Development, United Nations Development Programme, and United Nations Economic Commission for Africa. (2012). *African economic outlook 2012: Promoting youth employment.* OECD Publishing.

Agyeman, J. (2008). Toward a "just" sustainability? *Continuum, 22*(6), 751–756.

Agyeman, J. (2013). *Introducing just sustainabilities: Policy, planning, and practice,* Zed Books.

Ahmed, S., & Meenar, M. (2018). Just sustainability in the global south: A case study of the megacity of Dhaka. *Journal of Developing Societies, 34*(4), 1–24.

Ahwoi, K. (2010, April 26–27). *Government's role in attracting viable agricultural investment: Experiences from Ghana* [Conference presentation]. The World Bank Annual Bank Conference on Land Policy and Administration, Washington, DC, United States.

Akaabre, P. B., Poku-Boansi, M., Adarkwa, K. K. (2018). The growing activities of informal rental agents in the urban housing market of Kumasi, Ghana. *Cities, 83*(December), 34–43.

Akiwumi, F. (2017). Cultural conundrums in African land governance: Agribusiness in Sierra Leone. *Geography Research Forum, 37,* 37–60.

Alagidede, P., & Akpoza, A. (2015). Sovereign wealth funds and oil discovery: Lessons for Ghana, an emerging oil exporter. *Journal Africa Growth Agenda, 10,* 7–10.

Alchian, A. A., & Demsetz, H. (1973). The property right paradigm. *The Journal of Economic History, 33*(1), 16–27.

Alexander, G. S., & Peñalver, E. M. (2012). *An introduction to property theory.* Cambridge University Press.

Aligica, P. D., & Tarko, V. (2012). Polycentricity: From Polanyi to Ostrom and beyond. *Governance, 25*(2), 237–262.

Allan, T. (2003). Virtual water – The water, food, and trade nexus useful concept or misleading metaphor? *Water International, 28*(1), 4–11.

Allan, T., Keulertz, M., Sojamo, S., & Jeroen, W. (Eds.). (2013). *Handbook of land and water grabs in Africa.* Routledge.

Allen, D. W. E., & Potts, J. (2016). How innovation commons contribute to discovering and developing new technologies. *International Journal of the Commons, 10*(2), 1035–1054.

Amadae, S. M. (2004). Bargaining with the devil: Commentary on the Ostrom's "Quest for Meaning in Public Choice." *American Journal of Economics and Sociology, 63*(1), 161–165. https://doi.org/10.1111/j.1536-7150.2004.00279.x

Amadae, S. M. (2015). *Prisoners of reason: Game theory and neoliberal political economy.* Cambridge University Press.

Amankwah-Amoah, J., & Osabutey, E. (2018). Newly independent nations and large engineering projects: The case of the Volta River Project. *Critical Perspectives on International Business, 14*(2/3), 154–169.

Amanor-Wilks, D. (2009). Land, labour and gendered livelihoods in a "peasant" and a "settler" economy. *Feminist Africa, 12,* 31–50.

Amao, O. O. (2008). Corporate social responsibility, multinational corporations and the law in Nigeria: Controlling multinationals in host states. *Journal of African Law, 52*(1), 89–113.

Amin, S. (1990). *Delinking: Towards a polycentric world.* Zed Books.

Amin, S. (2014). Understanding the political economy of contemporary Africa. *Africa Development, 39*(1), 15–36.

Amin, A., & Howell, P. (2016). *Releasing the commons: Rethinking the futures of the commons,* Routledge.

Amnesty International. (2015). *Too toxic to touch? The UK's response to Amnesty International's call for a criminal investigation into Trafigura Ltd.*

Amoah, L. G. A. (2014). China, architecture and Ghana's spaces: Concrete signs of a soft Chinese imperium? *Journal of Asian and African Studies, 51*(2), 238–255.

Amoako-Tuffour, J. (2016). Should countries invest resource revenues abroad when demands for public infrastructure are pressing at home? The dilemma of sovereign wealth funds in Sub-Saharan Africa. *Journal of African Economies,* 25(Suppl. 2), ii41–ii58.

Anderson, A. (2015a). Garbage in, garbage out. *The Economist.*

Anderson, B. (2015b). Marcelo Diverse and Claudio Moreira: Betweener talk: Decolonizing knowledge production, pedagogy, and praxis. *Journal of Economic Issues, 49*(3), 891–893.

Anderson, T. (2011). Melanesian land: The impact of markets and modernisation. *Journal of Australian Political Economy, 68,* 86–107.

Anderson, T. L., & Libecap, G. D. (2014). *Environmental markets: A property rights approach.* Cambridge University Press.

Andrés, A. R., & Asongu, A. S. (2019). Trajectories of knowledge economy in SSA and MENA countries. *Technology in Society.* https://doi.org/10.1016/j.techsoc.2019.03.002

Andrianisa, H. A., Brou, Y. O. K., & Séhi bi, A. (2016). Role and importance of informal collectors in the municipal waste pre-collection system in Abidjan, Côte d'Ivoire. *Habitat International, 53,* 265–273.

Appessika, K. (2003). *Understanding slums: The case of Abidjan, Ivory Coast.* University College London. https://www.ucl.ac.uk/dpu-projects/Global_Report/pdfs/Abidjan.pdf

Apter, A. (2005). *The pan-African nation: Oil and the spectacle of culture in Nigeria.* University of Chicago Press.

Arezki, R., Deininger, K., & Selod, H. (2015). What drives the global "land rush"? *The World Bank Economic Review, 29*(2), 207–233.

Argyrous, G. (2017). Cost-benefit analysis as operationalized neoclassical economics: from evidence to folklore. *Journal of Australian Political Economy, 80*, 201–211.

Arias, M., Atienza, M., & Cademartori, J. (2014). Large mining enterprises and regional development in Chile: between the enclave and cluster. *Journal of Economic Geography, 14*, 73–95.

Arku, G., Yeboah, I. E. A., & Nyantakyi-Frimpong, H. (2016). Public parks as an element of urban planning: A missing piece in Accra's growth and development. *Local Environment, 21*(12). https://doi.org/10.1080/13549839.2016.1140132

Asante, L. A. (2020). Urban governance in Ghana: the participation of traders in the redevelopment of Kotokuraba Market in Cape Coast. *African Geographical Review.* https://doi.org/10.1080/19376812.2020.1726193

Asante, L. A., & Helbrecht, I. (2018). Seeing through African protest logics: A longitudinal review of continuity and change in protests in Ghana. *Canadian Journal of African Studies / Revue canadienne des études africaines, 52*(2), 159–181.

Asante, S. K. B. (1965). Interests in land in the customary law of Ghana: A new appraisal. *The Yale Law Journal, 74*(5), 848–885.

Asante, S. K. B. (1975). *Property law and social goals in Ghana, 1844–1966.* Ghana Universities Press.

Asongu, S., & Nwachukwu, J. C. (2018). *The mobile phone as an instrument of reducing information asymmetry for financial efficiency.* AGDI Working Paper. AfricanGovernance and Development Institute.

Aurigi, A., & Odendaal, N. (2020). From "smart in the box" to "smart in the city": Rethinking the socially sustainable smart city in context. *Journal of Urban Technology.* https://doi.org/10.1080/10630732.2019.1704203

Austin, G. (2005)., *Labour, land and capital in Ghana: From slavery to free labour in Asante, 1807–1956.* University of Rochester Press.

Awuah-Nyamekye, S. (2013). *Managing the environmental crisis in Ghana: The role of African traditional religion and culture—A case study of Berekum traditional area* [Doctoral dissertation, School of Philosophy, Religion and the History of Science, University of Leeds]. CORE Depository. https://core.ac.uk/download/pdf/20077819.pdf

Axelsson, L. (2012). *Making borders: Engaging the threat of Chinese textiles in Ghana* [Doctoral dissertation, Department of Human Geography, Stockholm University]. DiVA Portal. http://www.diva-portal.org/smash/get/diva2:551083/FULLTEXT01.pdf

Balcom, P., & Carey, V. P. (2020). Exergy-based sustainability analysis for tile production from waste plastics in Uganda. *Journal of Energy Resources Technology, 142*(5), 050905. https://doi.org/10.1115/1.4045540

Balthrop, A. T. (2012). *Oil and gas production: An empirical investigation of the common pool* [Doctoral dissertation, Department of Economics, Georgia State University]. ScholarWorks. https://scholarworks.gsu.edu/econ_diss/80

Barbier, B. E. (2005). *Natural resources and economic development*. Cambridge University Press.

Barnett, V. (2004). The Russian "Obshchina" as an economic institution. *Journal of Economic Issues, 38*(4), 1037–1039.

Barrionuevo, N., & Peters, S. (2019). Against all odds: Oil culture and the commodity consensus in Argentinean Patagonia. In H. Graves & D. E. Beard (Eds.), *The rhetoric and discourse of oil* (Chapter 9). Taylor & Francis.

Bassey, N. (2012). *To cook a continent: Sestructive extraction and the climate crisis in Africa*. Pambazuka Press.

Bateman, M., Duvendack, M., & Loubere, N. (2019). Is fin-tech the new panacea for poverty alleviation and local development? Contesting Suri and Jack's M-Pesa findings published in *Science*. *Review of African Political Economy, 46*(161), 480–495. https://doi.org/10.1080/03056244.2019.1614552

Baumol, W. J. (2004). On entrepreneurship, growth, and rent-seeking: Henry George updated." *The American Economist, 48*(1), 9–16.

Bauwens, M., Kostakis, V., & Pasaitis, A. (2019). *Peer to peer: The commons manifesto*. Westminster University Press.

Bayat, A. (1997). *Poor people's movements in Iran: Street politics*. Columbia University Press.

Bayliss, K. (2014). The financialization of water. *Review of Radical Political Economics, 46*(3), 292–307.

Beauchemin, C., & Bocquier, P. (2004). Migration and urbanisation in francophone West Africa: An overview of the recent empirical evidence. *Urban Studies, 41*(11), 2245–2272.

Behdad, S. (1989). Property rights in contemporary Islamic economic thought: A critical perspective. *Review of Social Economy, 47*(2), 185–211.

Behrens, K., Kanemoto, Y., & Murata, Y. (2015). The Henry George Theorem in a second-best world. *Journal of Urban Economics, 85*, 34–51.

Bell, J., Huber, J., & Viscusi, W. K. (2016). *Fostering recycling participation in Wisconsin households through single-stream programs*. Vanderbilt Law and Economics Research Paper No. 16-3 Vanderbilt Public Law Research Paper No. 16-22. http://ssrn.com/abstract=2731294.

Bell, J., Huber, J., & Viscusi, W. K. (2017). Fostering recycling participation in Wisconsin households through single-stream programs. *Land Economics, 93*(3), 481–502.

Bell, S. A., Henry, J. F., & Wray, L. R. (2004). A chartalist critique of John Locke's theory of property, accumulation, and money: Or is it moral to trade your nuts for gold? *Review of Social Economy, 62*(1), 51–65.

Bennett N. J., Govan, H., & Satterfield, T. (2015). Ocean grabbing. *Marine Policy, 57*, 61–68.

Berry, S. (2001). *Chiefs know their boundaries: Essays on property, power, and the past in Asante, 1896–1996.* James Currey.

Béteille, A. (1998). The idea of Indigenous people. *Current Anthropology, 39*(2), 187–192.

Bhan, G. (2014). The real lives of urban fantasies. *Environment and Urbanization, 26*(1), 232–235.

Bible Study Tools. (n.d.). *Leviticus 25:23–28, New International Version.* Retrieved 7 April 2016 from https://www.biblestudytools.com/leviticus/ 25-23.html

Bieler, A., & Lee, C.-Y. (2017). Exploitation and resistance: A comparative analysis of the Chinese cheap labour electronics and high-value added IT sectors. In A. Bieler & L. Chun-Yi (Eds.), *Chinese labour in the global economy: Capitalist exploitation and strategies of resistance* (pp. 24–37). Routledge.

Birkeland, J. (2020). *Net-positive design and sustainable urban development.* Routledge.

Bish, R. L., & Ostrom, V. (1976). Understanding urban government. In H. M. Hochman (Ed.), *The urban economy* (pp. 95–117). W. W. Norton and Company.

Blaauw, D., Pretorius, A., & Schenck, R. (2019). The economics of urban waste picking in Pretoria. *African Review of Economics and Finance, 11*(2), 129–164.

Boakye, P. A., & Béland, D. (2019). Explaining chieftaincy conflict using historical institutionalism: A case study of the Ga Mashie chieftaincy conflict in Ghana. *African Studies, 78*(3), 403–422.

Boamah, F. (2014a). How and why chiefs formalise land use in recent times: the politics of land dispossession through biofuels investments in Ghana. *Review of African Political Economy, 41*(141), 406–423.

Boamah, F. (2014b). Imageries of the contested concepts "land grabbing" and "land transactions": Implications for biofuels investments in Ghana. *Geoforum, 54*, 324–334.

Bob-Milliar, G. M. (2019). "We run for the crumbs and not for office": The Nkrumahist minor parties and party patronage in Ghana. *Commonwealth & Comparative Politics, 57*(4), 445–465.

Bob-Milliar, G. M. (2020). Introduction: Methodologies for researching Africa. *African Affairs,* Article adaa011. https://doi.org/10.1093/afraf/adaa011

Bob-Milliar, G., & Obeng-Odoom, F. (2011). The informal economy is an employer, a nuisance, and a goldmine: Multiple representations of and responses to informality in Accra, Ghana. *Urban Anthropology and Studies of Cultural Systems and World Economic Development, 40*(3–4), 263–284.

Boettke, P. (2010). Is the only form of "reasonable regulation" self-regulation? Lessons from Lin Ostrom on regulating the commons and cultivating citizens. *Public Choice, 143*, 283–291.

Boettke, P. J., Fink, A., & Smith, D. J. (2012). The impact of Nobel Prize winners in economics: Mainline vs. mainstream economics. *American Journal of Economics and Sociology, 71*(5), 1219–1249.

Bollier, D., Helfrich, S., & The Heinrich Böll Foundation (Eds.). (2015). *Patterns of commoning.* The Commons Strategies Group.

Bolognesi, T., & Nahrath, S. (2020). Environmental governance dynamics: Some micro foundations of macro failures. *Ecological Economics, 170*, Article 106555. https://doi.org/10.1016/j.ecolecon.2019.106555

Bond, P. (2010). Water, health and the commodification debate. *Review of Radical Political Economics, 42*(4), 445–464.

Boonjubun, C., Vuolteenaho, J., & Haila, A. (2021). Religious land as commons: Buddhist temples, monastic landlordism, and the urban poor in Thailand. *American Journal of Economics and Sociology.*

Borch, C., & Kornberger, M. (2015). *Urban Commons,* Routledge.

Borowy, I., & Schmelzer, M. (Eds.). (2017). *History of the future of economic growth: Historical roots of current debates on sustainable degrowth.* Routledge.

Boyd, D. R. (2012). *The environmental rights revolution: A global study of constitutions, human rights, and the environment.* UBC Press.

Boydell, S. (2010). South Pacific Land: An alternative perspective on tenure traditions, business, and conflict. *Georgetown Journal of International Affairs, 11*(1), 17–25.

Boydell, S., & Searle, G. (2014). Understanding property rights in the contemporary urban commons. *Urban Policy and Research, 32*(3), 323–340.

Branch, A., & Mampilly, Z. (2015). *Africa uprising: Popular protest and political change.* Zed Books.

Bratton, M., & de Walle, N. V. (1994). Neopatrimonial regimes and political transitions in Africa. *World Politics, 46*(4), 453–489.

Brechbühl, S. (2011). *Female waste pickers in Côte d'Ivoire: A study of women's livelihoods in the informal waste management sector of Abidjan* [Master's thesis, Faculty of Natural Sciences, University of Berne]. Plateforme Re-Sources Documentation Centre. http://documents.plateforme-re-sources.org/wp-content/uploads/2016/01/A60-BRECHBUHL-Female-waste-pickers-in-C%23U00f4te-dIvoire.pdf

Brempong, A. (2007). *Transformations in traditional rule in Ghana (1951–1996).* Institute of African Studies.

Breul, M. (2019). Cities in "multiple globalizations": Insights from the upstream oil and gas World City Network. *Regional Studies, Regional Science, 6*(1), 25–31. https://doi.org/10.1080/21681376.(2018).1564628

Briefing: The new face of Facebook. (2016, April 9). *The Economist,* 18–21.

Bromley, D. W. (1991). *Environment and economy: Property rights and public policy.* Blackwell.

Bromley, D. W. (1992). The commons, common property, and environmental policy. *Environmental and Resougce Economics, 2*(1), 1–17.

Bromley, D. W. (2008). Resource degradation in the African commons: Accounting for institutional decay. *Environmental and Development Economics, 13*(5), 539–563.

Bromley, D. W. (2019). *Possessive individualism: A crisis of capitalism.* Oxford University Press.

Brown, D., McGranahan, G., & Dodman, D. (2014). *Urban informality and building a more inclusive, resilient and green economy.* IIED Working Paper. https://pubs.iied.org/pdfs/10722IIED.pdf

Brueckner, M., Durey, A., Mayes, R., & Pforr, C. (Eds.). (2014). *Resource curse or cure? On the sustainability of development in Western Australia.* Springer.

Bruncevic, M. (2017). *Space, materiality and the normative: Law, art and the commons.* Routledge.

Bryant, G. (2019). *Carbon markets in a climate-changing capitalism.* Cambridge University Press.

Bryceson, D. (2016). Beyond livelihoods: Occupationality and career formation in African artisanal mining. In K. Havnevik, T. Oestigaard, E. Tobisson, & T. Virtanen (Eds.), *Framing African development: Challenging concepts* (pp. 90–110). Brill.

Bryson, P. J. (2011). *The economics of Henry George: History's rehabilitation of America's greatest early economist.* Palgrave Macmillan.

Buchanan, J. M. (1965). An economic theory of clubs. *Economica, New Series, 32*(125), 1–14.

Burgin, A. (2013). Age of certainty: Galbraith, Friedman, and the public life of economic ideas. *History of Political Economy, 45,* 192–219.

Butler, G., Jones, E., & Stilwell, F. (2009). *Political economy now: The struggle for alternative economics at the University of Sydney.* Sydney University Press.

Butler, G., Jones, E., & Stilwell, F. (2009). *Political economy now! The struggle for alternative economics at the University of Sydney.* Darlington Press.

Cahill, D., Edwards, L., & Stilwell, F. (2012). Introduction: Understanding neoliberalism beyond the free market. In L. Edwards, D. Cahill, and F. Stilwell (Eds.), *Neoliberalism beyond the free market* (pp. 1–9). Edward Elgard Publishing.

Cahill, K., & McManon, R. (2010). *Who owns the world: The surprising truth about every piece of land on the planet.* Grand Central Publishing.

Cain, A. (2014). African urban fantasies: Past lessons and emerging realities. *Environment and Urbanization, 26*(2), 561–567.

Campaign against Climate Change. (2014). *One million climate jobs: Tackling the environmental and economic crises.* Marstan Press.

Campbell, P. F. (2014). *"The shack becomes the house, the slum becomes the suburb and the slum dweller becomes the citizen": Experiencing abandon and seeking legitimacy in Dar es Salaam* [Doctoral dissertation, School of Geographical and Earth Sciences College of Science and Engineering, University of Glasgow]. Enlighten: Theses. http://theses.gla.ac.uk/5612/

Carmody, P. (2011). *The new scramble for Africa.* Polity Press.

Castells, M. (1977). *The urban question: A Marxist approach.* Edward Arnold.

Castells, M. (1989). *The informational city*. Blackwell.

Castells, M. (2010). Globalisation, networking, urbanisation: Reflections on the spatial dynamics of the information age. *Urban Studies, 47*(13), 2737–2745.

Castree, N. (2008a). Neoliberasing nature: The logics of deregulation and reregulation. *Environment and Planning A, 40*, 131–152.

Castree, N. (2008b). Neoliberalizing nature: Processes, effects, and evaluations. *Environment and Planning A, 40*, 153–173.

Cato, M. S., & North, P. (2016). Rethinking the factors of production for a world of common ownership and sustainability: Europe and Latin America compared. *Review of Radical Political Economics, 48*(1), 36–52.

Cepek, M. L. (2018). *Life in oil: Cofan survival in the petroleum fields of Amazonia*. University of Texas Press.

Chakrabarti, A., Dhar, A., & Kayatekin, S. A. (2016). Editors' introduction. *Rethinking Marxism, 28*(3–4), 339–353.

Chang, H.-J. (2002). *Kicking away the ladder – Development strategy in historical perspective*. Anthem Press.

Chang, H.-J. (2011). Reply to the comments on "Institutions and economic development: Theory, policy and history." *Journal of Institutional Economics, 7*(4), 595–613.

Chen, Y., & Puttitanun, T. (2005). Intellectual property rights and innovation in developing countries. *Journal of Development Economics, 78*(2), 474–493.

Chen, Y., Salike, N., Luan, F., & He, M. (2016). Heterogeneous effects of inter – and intra-city transportation infrastructure on economic growth: Evidence from Chinese cities. *Cambridge Journal of Regions, Economy and Society, 9*, 571–587.

Chiweshe, M. (2017). Zimbabwe's land question in the context of large-scale land based investments. *Geography Research Forum, 37*, 13–36.

Chiweshe, M. K., & Chabata, T. (2019). The complexity of farmworkers' livelihoods in Zimbabwe after the fast track land reform: Experiences from a farm in Chinhoyi, Zimbabwe. *Review of African Political Economy, 46*(159), 55–70.

Christophers, B. (2018). *The new enclosure: The appropriation of public land in neoliberal Britain*. Verso.

Churches, L. (2009). Resource rent debate. *Progress, 1094*, 4–5. Prosper Australia. https://www.prosper.org.au/wp-content/uploads/Progress DEC2009_Final.pdf

Ciriacy-Wantrup, S. V., & Bishop, R. C. (1975). "Common property" as a concept in natural resource policy. *Natural Resource Journal, 15*, 713–727.

Cissé, D. (2013). South-South migration and Sino-African small traders: A comparative study of Chinese in Sengal and Africans in China. *African Review of Economics and Finance, 5*(1), 21–35.

Clay, N. (2017). Integrating livelihoods approaches with research on development and climate change adaptation. *Progress in Development Studies, 18*(1), 1–17.

Cleaver, F. (2002). Reinventing institutions: Bricolage and the social embeddedness of natural resource management. *The European Journal of Development Research, 14*(2), 11–30.

Cleaver, F. (2012). *Development through bricolage: Rethinking institutions for natural resource management.* Routledge.

Cleaver, F., & de Koning, J. (2015). Furthering critical institutionalism. *International Journal of the Commons, 9*(1), 1–18.

Cleveland, M. M. (2012). The economics of Henry George: A review essay. *American Journal of Economics and Sociology, 71*(2), 498–511.

The climate issue. (2019, September 21). *The Economist*, 11–12.

Cline-Cole, R. (2020). Bouquets and brickbats along the road to development freedom and sovereignty: Commentary on "Rethinking the idea of independent development and self-reliance in Africa." *African Review of Economics and Finance, 12*(1), 260–281.

Coase, R. H. (1960). The problem of social cost. *The Journal of Law & Economics, 3*, 1–44.

Cobb, C. W. (2000). *Measurement tools and the quality of life.* Redefining Progress.

Cobb, C. W. (2015). Competing theories of economic crisis. *American Journal of Economics and Sociology, 74*(2), 187–208.

Cobb, C. W. (2016). Questioning the commons: Power, equity, and the meaning of ownership. *American Journal of Economics and Sociology, 75*(2), 265–288.

Cobb, C. W. (2019). Editor's introduction: The social problem of monopoly. *American Journal of Economics and Sociology, 78*(5), 1043–1069.

Cobban, T. W. (2013). *Cities of oil: Municipalities and petroleum manufacturing in southern Ontario, 1860–1960.* University of Toronto Press.

Collier, P. (2008). *The bottom billion: Why the poorest countries are failing and what can be done about it.* Oxford University Press.

Collier, P. (2009). *Wars, guns and votes: Democracy in dangerous places.* Random House.

Collier, P. (2010). *The plundered planet: Why we must – and how we can – manage nature for global prosperity.* Oxford University Press.

Collier, P., & Venables, A. (2012). Greening Africa? Technologies, endowments and the latecomer effect. *Energy Economics, 34*, S75–S84.

Commons, J. R. (1924). *Legal foundations of capitalism.* Macmillan Company.

Connell, D., & Grafton, R. Q. (2011). Water reform in the Murray-Darling Basin. *Water Resources Research, 47*, 1–9.

Connell, R. (2007). *Southern theory: The global dynamics of knowledge in social science.* Allen and Unwin.

Convery, F., McDonnell, S., & Ferreira, S. (2007). The most popular tax in Europe? Lessons from the Irish plastic bags levy. *Environmental Resource Economics, 38*, 1–11.

Cooney, P., & Freslon, W. S. (2018). Introduction. *Research in Political Economy, 33*, 1–8.

Cooper, F. (2014). *Africa in the world: Capitalism, empire, nation-state*. Harvard University Press.

Côté-Roy, L., & Moser, S. (2019). Does Africa not deserve shiny new cities? The power of seductive rhetoric around new cities in Africa. *Urban Studies, 56*(12), 2391–2407.

Cousins, B. (2007). More than socially embedded: The distinctive character of "communcal tenure" regimes in South Africa and its implications for land policy. *Journal of Agrarian Change, 7*(3), 281–315.

Cox, O. C. (1945). An American dilemma: A mystical approach to the study of race relations. *Journal of Negro Education, 14*(2), 132–148.

Crenshaw, K. (1991). Mapping the margins: Intersectionality, identity politics, and violence against women of color. *Stanford Law Review, 43*, 1241–1299.

Cui, Z. (2011). Partial intimations of the coming whole: The Chongqing, experimenting light of the theories of Henry George, James Meade, and Antonio Gramsci. *Modern China, 37*(6), 646–660.

Curtis, M. (2016). *The new colonialism: Britain's scramble for Africa's energy and mineral resources*. War on Want.

Dagdeviren, H., & Robertson, S. A. (2013). A critical asssessment of incomplete contracts theory for private participation in public services: The case of water sector in Ghana. *Cambridge Journal of Economics, 37*(5), 1057–1075.

Dagdeviren, H., & Robertson, S. A. (2014). Political economy of privatization contracts: The case of water and sanitation in Ghana and Argentina. *Competition and Change, 18*(2), 150–163.

Dagdeviren, H., & Robertson, S. A. (2016). A critical assessment of transaction cost theory and governance of public services with special reference to water and sanitation. *Cambridge Journal of Economics, 40*(6), 1707–1724.

Daly, H. E., Cobb, J. B. Jr, & Cobb, C. W. (1994). *For the common good: Redirecting the economy toward community, the environment, and a sustainable future*. Beacon Press.

The danger of urban oil drilling. (2015, November 27). *The New York Times*. https://www.nytimes.com/2015/11/28/opinion/the-danger-of-urban-oil-drilling.html

A dangerous gap: The markets v. the real economy. (2020, May 7). *The Economist*. https://www.economist.com/leaders/2020/05/07/the-market-v-the-real-economy

Darity, W. A. Jr, & Hamilton, D. (2015). A tour de force in understanding intergroup inequality: An introduction to stratification economics. *Review of Black Political Economy, 42*(1–2), 1–6.

Darity, W. Jr. (2009). Stratification economics: Context versus culture and the reparations controversy. *Kansas Law Review, 57*, 795–811.

Date-Bah, S. K. (2015). *Reflections on the Supreme Court of Ghana*. Wildy Simmonds; Hill Publishing.

Datta, A. (2018). The digital turn in postcolonial urbanism: Smart citizenship in the making of India's 100 smart cities. *Transactions of the Institute of British Geographers, 43*, 405–419.

Davidson, D. J., & Dunlap, R. E. (2012). Introduction: Building on the legacy contributions of William R. Freudenburg in environmental studies and sociology. *Journal of Environmental Studies and Sciences, 2*(1), 1–6.

Dawson, A. (2016). *Extinction: A radical history*. OR Books.

Decker, F. (2015). Property ownership and money: A new synthesis. *Journal of Economic Issues, 49*(4), 922–946.

Deininger, K. (2003). *Land policies for growth and poverty reduction*. The World Bank.

de Koning, J. (2011). *Reshaping institutions – Bricolage processes in smallholder forestry in the Amazon* [Doctoral dissertation, Wageningen University]. Wageningen UR E-depot. https://edepot.wur.nl/160232

de Koning, J. (2014). Unpredictable outcomes in forestry – Governance institutions. *Society and Natural Resources, 27*(4), 358–371.

Demsetz, H. (1967). Toward a theory of property rights. *The American Economic Review, 57*(2), 347–359.

Demsetz, H. (2002). Toward a theory of property rights II: The competition between private and collective ownership. *The Journal of Legal Studies, 31*, s653–s672.

Denman, D. R. (1978). *The place of property: A new recognition of the function and form of property rights in land*. Geographical Publications; William Cloes and Sons.

Desmarais-Tremblay, M. (2019). The normative problem of merit goods in perspective. *Forum for Social Economics, 48*(3), 219–247.

de Soto, H. (1989). *The other path*. Harper & Row.

de Soto, H. (2000). *The mystery of capital: Why capitalism triumphs in the West and fails everywhere else*. Bantam Press.

de Soto, H. (2004). Bringing capitalism to the masses. *Cato's Letter, 2*(3), 1–8.

de Soto, H. (2011). This land is your land: A conversation with Hernando de Soto. *World Policy Journal, 28*, 35–40.

Destroying the city to save it. (2019, March 9). *The Economist*, 32.

de Vroey, M. (1975). The transition from classical to neoclassical economics: A scientific revolution. *Journal of Economic Issues, 9*(3), 415–439.

Dinko, D. H., Yaro, J., & Kusimi, J. (2019). Political ecology and contours of vulnerability to water insecurity in semiarid north-eastern Ghana. *Journal of Asian and African Studies, 54*(2), 282–299.

Diop, C. A. (1967). *The African origin of civilization: Myth or reality*. Laurence Hill.

Diop, C. A. (1977). *The African origin of civilization: Myth or reality* (3rd ed.). Laurence Hill.

Dirlik, A. (2007). Global South: Predicament and promise. *The Global South, 1*(1 and 2), 12–23. https://muse.jhu.edu/article/398223/pdf

Djezou, W. B. (2014). Community-based forest management in Côte d'Ivoire: A theoretical investigation. *African Review of Economics and Finance, 6*(2), 1–21.

Dobb, M. (1946). *Studies in the development of capitalism.* Routledge; Kegan Paul.

Dobeson, A. (2019). Review of *The new enclosure: The appropriation of public land in neoliberal Britain. British Journal of Sociology, 70*(3), 1095–1097.

Domeher, D., Frimpong, J. M., & Appia, T. (2014). Adoption of financial innovation in the Ghanaian banking industry. *African Review of Economics and Finance, 6*(2), 88–114.

dos Santos, A. O., Svensson, G., & Padin, C. (2013). Indicators of sustainable business practices: Woolworths in South Africa. *Supply Chain Management: An International Journal, 18*(1), 104–108.

Doussard, M., & Gamal, A. (2016). The rise of wage theft laws: Can community–labor coalitions win victories in state houses? *Urban Affairs Review, 52*(5), 780–807.

Dovey, K. (2012). Informal urbanism and complex adaptive assemblage. *International Development Planning Review, 34*(4), 349–367.

Dragun, A. K. (2001). Common property resources. In P. O'Hara (Ed.), *Encyclopedia of political economy* (pp. 118–120). Routledge.

Drakakis-Smith, D. (1987). *The third world city* (1st ed.). Methuen and Co.

Duchrow, U., & Hinkelammert, F. J. (2004). *Property for people, not for profit.* Zed Books.

Dugger, W. M. (1980). Property rights, law, and John R. Commons. *Review of Social Economy, 38*(1), 41–53.

Dugger, W. M. (2016). Technology and property: Knowledge and the commons. *Review of Radical Political Economics, 48*(1), 111–126.

Dunn, B. (2017). Against neoliberalism as a concept. *Capital and Class, 41*(3), 435–454.

Duvendack, M., & Mader, P. (2019). Impact of financial inclusion in low- and middle-income countries: A systematic review of reviews. *Campbell Systematic Reviews.* https://doi.org/10.4073/csr.2019.2

Dyasi, H. M. (1985). Culture and the environment in Ghana. *Environmental Management, 9*(2), 97–104.

Ebner, A. (2015). Marketization: Theoretical reflections building on the perspectives of Polanyi and Habermas. *Review of Political Economy, 27*(3), 369–389.

EcoConServ Environmental Solutions. (2016, February). *Benban 1.8GW PV Solar Park, Egypt, strategic environmental and social assessment: Final report.* https://www.ebrd.com/documents/environment/esia-48213nts.pdf

Ehwi, R. J., Morrison, N., & Tyler, P. (2019). Gated communities and land administration challenges in Ghana: Reappraising the reasons why people move into gated communities, *Housing Studies*. https://doi.org/10.1080/02673037.2019.1702927

Elahi, K. Q., & Stilwell, F. (2013). Customary land tenure, neoclassical economics and conceptual bias. *Niugini Agrisaiens, 5*, pp. 28–29.

Elhardary, Y. A. E., & Obeng-Odoom, F. (2012). Conventions, changes and contradictions in land governance in Africa: The story of land grabbing in Sudan and Ghana. *Africa Today, 59*(2), 59–78.

Elinor Ostrom: Facts. (2009). [Nobel Prize lecture]. The Nobel Prize. http://www.nobelprize.org/nobel_prizes/economic-sciences/laureates/2009/ostrom-facts.html

Eljuri, E., & Johnston, D. (2014). Mexico's energy sector reform. *The Journal of World Energy and Business, 7*(2), 168–170.

Elsner, W., Heinrich, T., & Schwardt, H. (2014). *The microeconomics of complex economies: Evolutionary, institutional, neoclassical, and complexity perspectives.* Elsevier/Academic Press.

Ely, R. T. (1916). Russian land reform. *American Economic Review, 6*(1), 61–68.

Ely, R. T. (1917). Landed property as an economic concept and as a field of research. *The American Economic Review, 7*(Suppl. 1), 18–33.

Energy Commission. (2010). *Bioenergy policy for Ghana.* http://cleancookstoves.org/resources_files/draft-bioenergy-policy-for.pdf

Engels, F. (1928). *The mark.* Labor News Co. (Original work published 1892)

Entsua-Mensah, M. (2001, August 10–12). *Traditional management of water resources in West Africa* [Conference presentation]. International Water History Association's Conference on the Role of Water in History and Development, Bergen, Norway.

Erickson, G., & Groh, C. (2012). How the APF and the PFD operate: The peculiar mechanics of Alaska's state finances. In K. Widerquist & M. W. Howard (Eds.), *Alaska's permanent fund dividend: Examining its suitability as a model* (pp. 41–48). Palgrave Macmillan.

Eshun, E. K. (2011). *Religion and nature in Akan culture: A case study of Okyeman Environment Foundation* [Unpublished master's thesis]. Department of Religious Studies, Queen's University Kingston.

Eskom emissions bring early death to thousands. (2018, November 24). *Pretoria News*, 6.

Espin-Sanchez, J.-A. (2015). Review of "The commons in history: Culture, conflict and ecology." *Journal of Economic History, 75*(1), 270–272.

Esteva, G. (2014). Commoning in the new society. *Community Development Journal, 49*(S1), i144–i159.

Estevez, D. (2013, June 26). *Most Mexicans oppose President Peña Nieto's plans to open up Pemex to private investment.* Forbes. http://www.forbes.com/sites/

doliaestevez/2013/06/26/most-mexicans-oppose-president-pena-nietos-plans-to-open-up-pemex-to-private-investment/

Euler, J. (2016). Commons-creating society: On the radical German commons discourse. *Review of Radical Political Economics, 48*(1), 93–110.

Euler, J. (2018). The commons: A social form that allows for degrowth and sustainability. *Capitalism Nature Socialism.* https://doi.org/10.1080/10455 752.2018.1449874

Exner, A. (2014). Degrowth and demonetization: On the limits of a non-capitalist market. *Capitalism, Nature, Socialism,* 25(3), 9–27.

Exner, A. (2015). Commons: Eon nomadisierender Begriff im Wandel von Bedeutungsfeldern. Anmerkungen zur theoretischen Analyse des Werks von Elinor Ostrom und linksalternativer Bezüge darauf. *Emanzipation, 5*(1), 119–155. http://emanzipation.org/articles/em_5-1/e_5-1_exner.pdf

Exner, A. (2021). How to conceptualize the commons? In A. Exner, S. Kumnig, F. Obeng-Odoom, S. Hochleither (Eds.), *Whose commons? Appropriations, contestations, and perspectives.* Routledge.

Exner, A., Bartels, L. E., Windhaber, M., Fritz, S., See, L., Politti, E., & Hochleithner, S. (2015). Constructing landscapes of value: Capitalist investment for the acquisition of marginal or unused land—The case of Tanzania. *Land Use Policy, 42,* 652–663.

Fahnbulleh, M. (2020). The neoliberal collapse. *Foreign Affairs,* 99(1), 38–43.

Fairhead, J., Leach, M., & Scoones, I. (2012). Green grabbing: a new appropriation of nature? *The Journal of Peasant Studies, 39*(2), 237–261.

Fanon, F. (1961). *The wretched of the earth.* Grove Press.

Fine, B. (1990). *The coal question: Political economy and industrial change from the nineteenth century to the present day.* Routledge.

Fine, B. (2010a). *Theories of social capital: Researchers behaving badly.* Pluto.

Fine, B. (2010b). Beyond the tragedy of the commons: A discussion of *Governing the commons: The evolution of insitutions for collective action. Perspectives on Politics, 8*(2), 583–586.

Fine, B., & Rustomjee, Z. (1996). *The political economy of South Africa: From minerals-energy complex to industrialisation.* Westview Press.

Flanagan, F., & Stilwell, F. (2018). Causes and consequences of labour's falling income share and growing inequality. *Journal of Australian Political Economy, 81,* 5–10.

Fonjong, L. N., & Fokum, V. Y. (2015). Rethinking the water dimension of large scale land acquisitions in sub-Saharan Africa. *Journal of African Studies and Development, 7*(4), 112–120.

Forstater, M. (2005). Taxation and primitive accumulation: The case of colonial Africa. *Research in Political Economy, 22,* 51–64.

Foster, E. A. (2019). *African Catholic: Decolonization and the transformation of the church.* Harvard University Press.

Foster, S. R. (2016). The city as a commons. *Yale Law and Policy Review, 34*(2), 282–349.

Foster, S. R., & Iaione, C. (2019). Ostrom in the city: Design principles and practices for the urban commons. In B. Hudson, J. Rosenbloom, & D. Cole (Eds.), *Routledge handbook of the study of the commons* (pp. 235–255). Routledge.

Fosu, A. K., & Gafa, D. W. (2020). Progress on poverty in Africa: How have growth and inequality mattered? *African Review of Economics and Finance, 12*(1), 61–101.

Fox, M. H. (2014). *Why we need nuclear power: The environmental case.* Oxford University Press.

Fox, S. (2014). The political economy of slums: Theory and evidence from Sub-Saharan Africa. *World Development, 54,* 191–203.

Fox, S., Bloch, R., & Monroy, J. (2018). Understanding the dynamics of Nigeria's urban transition: A refutation of the "stalled urbanisation hypothesis." *Urban Studies, 55,* 947–964.

Francis, P. (2015). *On care for our common home: Encyclical letter of the Holy Father Francis.* Vatican Press.

Franco, J., Metha, L., & Veldwisch, G. J. (2013). The global politics of water grabbing. *Third World Quarterly, 34*(9), 1651–1675.

Francois, J. H. (1995). Forest resources management in Ghana. In Ghana Academy of Arts and Social Sciences (Ed.), *Management of Ghana's natural resources: Vol. 34* (pp. 121–131). Ghana Academy of Arts and Social Sciences.

Frankel, E. G. (2007). *Oil and security: A world beyond petroleum.* Springer.

Freedman, L., & Michaels, J. (2019). *The evolution of nuclear strategy* (Rev. 4th ed.). Palgrave Macmillan.

Friedman, T. L. (2007). *The world is flat: A brief history of the twenty-first century.* Picador.

Friedrich-Ebert Foundation. (2011). *Youth and oil and gas: Governance in Ghana nationwide survey.*

Frischmann, B. M. (2013). Two enduring lessons from Elinor Ostrom. *Journal of Institutional Economics, 9*(4), 387–406.

Fukuyama, F. (1989). The end of history? *The National Interest,* (Summer), 1–18.

Fukuyama, F. (1992). *The End of History and the Last Man,* The Free Press, New York.

Gaffney, M. (2008). Keeping land in capital theory: Ricardo, Faustmann, and George. *American Journal of Economics and Sociology, 67*(1), 119–142

Gaffney, M. (2015). A real-assets model of economic crises: Will China crash in 2015? *American Review of Economics and Sociology, 74*(2), 325–360.

Gaisie, E., Kim, H. M., & Han, S. S. (2019). Accra towards a city-region: Devolution, spatial development and urban challenges. *Cities, 95,* 102398.

Galbraith, J. K. (1977). *The age of uncertainty: A history of economic ideas and their consequences.* Houghton Mifflin Company.

Galbraith, J. K. (1998). *The affluent society.* Houghton Mifflin Company. (Original work published 1958)

Garnett Jr., R. F. (2019). Smith after Samuelson: Care and harm in a socially entangled world. *Forum for Social Economics, 48*(2), 125–136.

Ge, J., & Lei, Y. (2018). Resource tax on rare earths in China: Policy evolution and market responses. *Resources Policy, 59,* 291–297.

Gellers, J. C. (2012). Greening constitutions with environmental rights: Testing the isomorphism thesis. *Review of Policy Research, 29*(4), 523–543.

Gellers, J. C. (2015). Explaining the emergence of constitutional environmental rights: A global quantitative analysis. *Journal of Human Rights and the Environment, 6*(1), 75–97.

George, H. (1884). *Moses.* The United Committee for Taxation of Land Values.

George, H. (1885). *The crime of poverty.* History Is a Weapon. http://www .historyisaweapon.com/defcon1/georgecripov.html

George, H. (1891). *The condition of labor – An open letter to Pope Leo XIII.* Weath and Want. http://wealthandwant.com/pdf/George_The_Condition_of_ Labor.pdf

George, H. (1935). *Progress and poverty* (50th anniversary ed.). Robert Schalkenbach Foundation.

George, H. (1966). *Social problems.* Robert Schalkenbach Foundation. (Original work published 1883)

George, H. (1981). *A perplexed philosopher.* Robert Schalkenbach Foundation. http://schalkenbach.org/library/henry-george/perplexedphilosopher/ (Original work published (1892)

George, H. (1991). *Protection or free trade: An examination of the tariff question, with especial regard to the interests of labor.* Robert Schalkenbach Foundation. http://schalkenbach.org/library/henry-george/protection-or-free-trade/ preface-index.html (Original work published 1886)

George, H. (1992). *The science of political economy.* Kegan Paul, Trench, Trubner and Co.; Robert Schalkenbach Foundation. (Original work published 1898)

George, H. (2006). *Progress and poverty.* Robert Schalkenbach Foundation. (Original work published 1879)

Gerber, J.-D., Knoepfel, P., Nahrath, S., & Varone, F. (2009). Institutional Resource Regimes: Towards sustainability through the combination of property-rights theory and policy analysis. *Ecological Economics, 68,* 798–809.

Gërxhani, K. (2004). The informal sector in developed and less developed countries: A literature review. *Public Choice, 120,* 267–300.

Gibson-Graham, J. K. (2006a). *A postcapitalist politics.* University of Minnesota Press.

Gibson-Graham, J. K. (2006b). *The end of capitalism (as we knew it): A feminist critique of political economy.* University of Minnesota Press.

Gilbert, A. (2012). De Soto's *The mystery of capital*: Reflections on the book's public impact. *International Development Planning Review, 34*(3), v–xvii.

Giles, R. (2015). A review of "Nature's gifts." *Good Government, 1027*, 11–12.

Giles, R. (2016a, June 12). *A review of "Nature's gifts"* [Paper presentation]. Sydney, Australia.

Giles, R. (2016b). *The theory of charges on commonland.* Association for Good Government.

Giles, R. (2017). *The theory of charges for nature.* Association for Good Government.

Gills, B. K. (2016). Interview with Boris Kagarlitsky. *Third World Quarterly, 37*(4), 744–748.

Gómez-Baggethun, E., & Muradian, R. (2015). In markets we trust? Setting the boundaries of market-based instruments in ecosystem services governance. *Ecological Economics, 117*(September), 217–224.

Gonce, R. A. (1996). The social gospel, Ely, and commons's initial stage of thought. *Journal of Economic Issues, 30*(3), 641–665.

Goodman, J., & Rosewarne, S. (2011). Challenging climate change: Introduction. *Journal of Australian Political Economy, 66*(Summer), 5–16.

Goodman, J., & Rosewarne, S. (2015). Slowing uranium in Australia: Lessons for urgent transition beyond coal, gas, and oil. In T. Princen, P. Jack, P. Manno, & L. Martin (Eds.), *Ending the fossil fuel era* (pp. 193–222). MIT Press.

Goodman, J., & Worth, D. (2008). The minerals boom and Australia's resource curse. *Journal of Australian Political Economy, 61*, 201–219.

Gopaldas, R. (2016). Money goes online. *Africa in Fact, 38*(July/August), 55–58.

Gordon Nembhard, J. (2014a). Community asset building and community wealth. National Economic Association Presidential Address. *The Review of Black Political Economy 41*(2), 101–117.

Gordon Nembhard, J. (2014b). *Collective courage: A history of African American cooperative economic thought and practice.* Pennsylvania State University Press.

Gore, C. (1996). Methodological nationalism and the misunderstanding of East Asian industrialization. *European Journal of Development Research, 8*(1), 77–122.

Gore, C. (2017). Late industrialisation, urbanisation and the middle-income trap: An analytical approach and the case of Vietnam. *Cambridge Journal of Regions, Economy and Society, 10*(1), 35–58.

Gould, J. (2006). Strong bar, weak state? Lawyers, liberalism and state formation in Zambia. *Development and Change, 37*(4), 921–941.

Graeber, D. R. (2011). *Debt: The first 5000 years.* Melville House.

Grant, R. (2009). *Globalizing city: The urban and economic transformation of Accra, Ghana.* Syracuse University Press.

Grant, R. (2015). *Africa: Geographies of change.* Oxford University Press.

Grant, R., & Oteng-Ababio, M. (2016). The global transformation of materials and the emergence of informal urban mining in Accra, Ghana. *Africa Today, 62*(4), 2–20.

Grant, S. A. (1976). Obshchina and Mir. *Slavic Review, 35*(4), 636–651.

Green, R. H. (1987). *Stabilization and adjustment policies and programmes: Country study 1, Ghana*. World Institute for Development Economics Research; United Nations University.

Gronow, J. (1997). *The sociology of taste*. Routledge.

GSMA. (2014). *State of the industry: Mobile financial services for the unbanked*.

Guha-Khasnobis, B., Kanbur, R., & Ostrom, E. (Eds.). (2006). *Linking the formal and informal economy*, Oxford Universities Press.

Gunderson, R. (2018). Degrowth and other quiescent futures: Pioneering proponents of an idler society. *Journal of Cleaner Production, 198*, 1574–1582.

Gunn, C. (2015). Acequias as Commons: Lessons for a post-capitalist world. *Review of Radical Political Economics, 48*(1), 81–92

Gwama, M. (2014). Explaining weak financial development in Africa. *African Review of Economics and Finance, 6*(2), 69–88.

Gyampo, R., & Obeng-Odoom, F. (2013). Youth participation in local and national development in Ghana, 1620–2013. *Journal of Pan African Studies, 5*(9), 129–150.

Gyau-Boakye, P. (2001). Sources of rural water supply in Ghana. *Water International, 26*(1), 96–104.

Haila, A. (2000). Real estate in global cities: Singapore and Hong Kong as property states. *Urban Studies, 37*(12), 2241–2256.

Haila, A. (2011). The comedy of the suburban commons: A Chinese story. In E. Kahla (Ed.), *Between Utopia and Apocalypse: Essays on social theory and Russia* (pp. 103–112). Aleksanteri Institute.

Haila, A. (2016). *Urban land rent: Singapore as a property state*. Wiley Blackwell.

Haila, A. (2017). Institutionalizing "the property mind." *International Journal of Urban and Regional Research, 41*(3), 500–507.

Haila, A. (2018). *Urban land tenure* [Project document]. Academy of Finland.

Haller, T., Breu, T., de Moor, T., Rohr, C., & Znoj, H. (Eds.). (2019). *The commons in a glocal world: Global connections and local responses*. Routledge.

Hamel, P., & Keil, R. (Eds.). (2015). *Suburban governance: A global view*. University of Toronto Press.

Hardin, G. (1968). The tragedy of the commons. *Science, 162*(3859), 1243–1248.

Hardin, G. (1974, September). Lifeboat ethics: The case against helping the poor. *Psychology Today*, 800–812.

Hardus, S. (2014). Chinese national oil companies in Ghana: The cases of CNOOC and Sinopec. *Perspectives on Global Development and Technology, 13*(5–6), 588–612.

Hartwick, J. M. (1977). Intergenerational equity and the investing of rents from exhaustible resources. *The American Economic Review, 67*(5), 972–974.

Harvey, D. (2003). *The new imperialism*. Oxford University Press.

Harvey, D. (2011). The future of the commons. *Radical History Review, 109*(Winter), 101–107.

Harvey, D. (2012). *Rebel cities: From the right to the city to the urban revolution*. Verso.

Hayek, F. A. (1945). *The road to serfdom*. The Institute of Economic Affairs.

Head-König, A.-L. (2019). The commons in highland and lowland Switzerland over time: Transformations in their organisation and survival strategies (seventeenth to twentieth century). In T. Haller, T. Breu, T. de Moor, C. Rohr, & H. Znoj (Eds.), *The commons in a global world: Global connections and local responses* (pp. 156–172). Routledge.

Heffron, R. J., & McCauley, D. (2018). What is "just transition"? *Geoforum, 88*, 74–77.

Heideman, P. (2015, April 7). Technology and socialist strategy. *Jacobin*. https://jacobinmag.com/2015/04/braverman-gramsci-marx-technology

Hess, J. B. (2000). Imaging architecture: The structure of nationalism in Accra, Ghana. *Africa Today, 47*(2), 35–58.

Hickel, J. (2019). The contradiction of the sustainable development goals: Growth versus ecology on a finite planet. *Sustainable Development, 27*(5), 873–884.

Hiedanpää, J., & Bromley, D. W. (2016). *Environmental heresies*. Palgrave Macmillan.

Hill, P. (1966). A plea for Indigenous economics: The West African example. *Economic Development and Cultural Change, 15*(1), 10–20.

Hill, P. (1961). The migrant cocoa farmers of southern Ghana. *Africa: Journal of the International African Institute, 31*(3), 209–230.

Hillbom, E., & Bolt, J. (2018). *Botswana – A modern economic history: An African diamond in the rough*. Palgrave.

Hilson, A. E. (2014). *Resource enclavity and Corporate Social Responsibility in Sub-Saharan Africa: The case of oil production in Ghana* (Publication No. 24545) [Doctoral dissertation, Aston University]. Aston Publications Explorer. http://publications.aston.ac.uk/id/eprint/24545/1/Hilson_Abigail_2014.pdf

Hirschman, A. (1970). *Exit, voice, and loyalty: Responses to decline in firms, organizations, and states*. Harvard University Press.

Hodgson, G. M. (2013). Editorial introduction to the Elinor Ostrom memorial issue. *Journal of Institutional Economics, 9*(4), 381–385.

Hodgson, G. M. (2014). On fuzzy frontiers and fragmented foundations: some reflections on the original and new institutional economics. *Journal of Institutional Economics, 10*(4), 591–611

Holstenkamp, L. (2019). What do we know about cooperative sustainable electrification in the Global South? A synthesis of the literature and refined social-ecological systems framework. *Renewable and Sustainable Energy Reviews, 109*, 307–320.

hooks, b. (1982). *Ain't I a woman? Black women and feminism*. Pluto.

Hossein, C. S. (2016). *Politicized microfinance: Money, power, and violence in the black Americas*. University of Toronto Press.

Hossein, C. S. (Ed.). (2018). *The black social economy in the Americas: Exploring diverse community-based markets*. Palgrave Macmillan.

Hotelling, H. (1931). The economics of exhaustible resources. *The Journal of Political Economy, 39*(2), 137–175.

How to beat the big men. (2020, March 7). *The Economist,* 14.

Hubert, M. K. (1974, June 6). *Testimony to Hearing on the National Energy Conservation Policy Act of 1974, hearings before the Subcommittee on the Environment of the Committee on Interior and Insular Affairs House of Representatives.*

Hughes, R. B. (2016). The autonomous vehicle revolution and the global commons. *SAIS Review of International Affairs, 36*(2), 41–56.

Hujo, K. (Ed.). (2012). *Mineral rents and the financing of social policy: Opportunities and challenges.* Palgrave.

Huntington, S. P. (1993). The clash of civilizations? *Foreign Affairs, 72*(3), 22–49.

Huntington, S. P. (1996). *The clash of civilizations and the remaking of world order.* Simon & Schuster.

Huron A. (2018). *Carving out the commons: Tenant organizing and housing cooperatives in Washington, D.C.* University of Minnesota Press.

Hymer, H. S. (1970). Economic forms in pre-colonial Ghana. *The Journal of Economic History, 30*(1), 33–50.

Iaione, C. (2016). The co-city: Sharing, collaborating, cooperating, and commoning in the city. *American Journal of Economics and Sociology, 75*(2), 415–455.

IEA. (2011). The case for windfall profit taxes in Ghana's mining code. *Legislative Alert, 19,* 5.

Ince, O. U. (2014). Primitive accumulation, new enclosures, and global land grabs: A theoretical intervention. *Rural Sociology, 79*(1), 104–131.

Ingold, A. (2018). Commons and Environmental regulation in history: the water commons beyond property and sovereignty. *Theoretical Inquiries in Law, 19*(2), 425–456.

Issah, Z. (2013, April 21). Stop foreigners from taking over our lands – Varsity registrar pleads with Supreme Court. *Daily Graphic.*

Jacob, T. (2017). Competing energy narratives in Tanzania: Towards the political economy of coal. *African Affairs, 116*(463), 341–353.

Jacobs, H. M. (2020). Whose citizenship? Social conflict over property in the United States. *Social Policy and Society, 19*(2), 343–355.

Jacobs, H. M. (1995). The anti-environmental "wise use" movement in America. *Land Use Law &Zoning Digest, 47*(2), 3–8.

James, N. (2019). The coal debate. *Mining Weekly, 25*(3), 10–11.

Jamieson, J. (2014). Property pirates – Why are they given carte blanche? *Progress, 1112,* 17–19.

Jamison, A. (1998). American anxieties: Technology and the reshaping of republican values. In M. Hård & A. Jamison (Eds.), *The intellectual appropriation of technology: Discourses on modernity, 1900–1939* (pp. 69–100). MIT Press.

Jang, H. S. (2015). *Social identities of young Indigenous people in contemporary Australia: Neo-colonial North, Yarrabah.* Springer.

Jeffords, C. (2015). *A panel data analysis of the effects of constitutional environmental rights provisions on access to improved sanitation facilities and water source.* Economic Rights Working Papers 24. University of Connecticut; Human Rights Institute. https://ideas.repec.org/p/uct/ecriwp/hri24.html

Jeffords, C., & Minkler, L. (2014). *Do constitutions matter? The effects of constitutional environmental rights provisions on environmental outcomes.* Working papers 2014-16. University of Connecticut, Department of Economics. https://ideas.repec.org/p/uct/uconnp/2014-16.html

Jeffords, C., & Minkler, L. (2016). Do constitutions matter? The effects of constitutional environmental rights provisions on environmental outcomes. *KYKLOS, 69*(2), 294–335.

Jeffords, C., & Shah, F. (2013). On the natural and economic difficulties to fulfilling the human right to water within a neoclassical economics framework. *Review of Social Economy, 71*(1), 65–92.

Jerven, M. (2015). *Africa: Why economists get it wrong.* Zed Books.

Jevons, W. S. (2012). *The coal question: An inquiry concerning the progress of the nation, and the probable exhaustion of our coal-mines.* Forgotten Books; Macmillan and Co. (Original work published 1906)

Jibao, S. S., & Prichard, W. (2015). The political economy of property tax in Africa: Explaining reform outcomes in Sierra Leone. *African Affairs, 114*(456), 404–431.

Johnstone, N., & Wood, L. (1999). *Private sector participation in water supply and sanitation: Realising social and environmental objectives.* International Institute for Environment and Development.

Jones, E. (2010). The Chicago School, Hayek and the Mont Pélerin Society. *Journal of Australian Political Economy, 35*(Winter), 139–155.

Judicial Service of Ghana. (2006) Tribute by Judicial Service. In M. Abakah, *Celebration of life: Mr. Justice Moses Abakah, retired supervising high court judge, a.k.a. Kofi Baah-Kofi Boye* (pp. 18–19). Max Creation.

Judis, J. B. (2013). Marx is dead, long live Marx's ideas. *Dissent, 61*(1), 76–80.

Kahneman, D. (2011). *Thinking, fast and slow.* Penguin Books; Random House.

Kaplan, F. (2016). Rethinking nuclear policy: Taking stock of the stockpile. *Foreign Affairs, 95*(5), 18–25.

Kapp, K. W. (1971). *The social costs of private enterprise.* Schocken Books.

Karikari, I. (2006, February). *Ghana's Land Administration Project (LAP) and Land Information Systems (LIS) implementation: The issues* [Paper presentation]. International Federation of Surveyors. https://www.fig.net/resources/monthly_articles/2006/february_2006/karikari_february_2006.pdf

Karikari, I., Stillwell, J., & Carver, S. (2003). Land administration and GIS: The case of Ghana. *Progress in Development Studies, 33*, 223–242.

Kea, P. (2013). The complexity of an enduring relationship: Gender, generation and the moral economy of the Gambian Mandinka household. *Journal of the Royal Anthropological Institute, 19*(1), 102–119.

Kea, P. J. (2010). *Land, labour and entrustment.* Brill.

Keen, S. (2003). Madness in their method. In F. Stilwell & G. Argyrous (Eds.), *Economics as a Social Science* (pp. 140–145). Pluto Press.

Kelly, J. M. (1981). The new barbarians: The continuing relevance of Henry George. *American Journal of Economics and Sociology, 40*(3), 299–308.

Kepe, T. (2008). Land claims and co-management of protected areas in South Africa: Exploring the challenges. *Environmental Management, 41*(3), 311–321.

Keynes, J. M. (1936). William Stanley Jevons, 1835–1882: A centenary allocation on his life and work as economist and statistician. *Journal of the Royal Statistical Society, 99*(3), 516–555.

Khan, M. E. (2010). *Climatopolis: How Our cities will thrive in the hotter future.* Basic Group.

Kim, J., & Mahoney, J. T. (2002). Resource-based and property rights perspectives on value creation: The case of oil field unitization. *Managerial and Decision Economics, 23,* 225–245.

Kimball, A. (1973). The first international and the Russian Obshchina. *Slavic Review, 32*(3), 491–514.

Kitchin, R. (2015). Making sense of smart cities: Addressing present shortcomings. *Cambridge Journal of Regions, Economy and Society, 8,* 131–136.

Kizito, F., Williams, T. O., McCartner, M., & Erkossa, T. (2013). Green and blue water dimensions of foreign direct investment in biofuel and food production in West Africa: The case of Ghana and Mali. In T. Allan, M. Keulertz, S. Sojamo, & J. Warner (Eds.), *Handbook of land and water grabs in Africa: Foreign direct investment and food and water security* (pp. 337–358). Routledge.

Klaes, M. (2000). The birth of the concept of transaction costs: Issues and controversies. *Industrial and Corporate Change, 9*(4), 567–593.

Klossek, P., Kullik, J., & van den Boogaart, K. G. (2016). A systemic approach to the problems of the rare earth market. *Resources Policy, 50,* 131–140.

Komenan, N. (2010, October). *Water and sanitation in Côte d'Ivoire: Before and after crisis.* Working paper. University of Cocody.

Konadu-Agyemang, K. (2000). *The political economy of housing and urbanization in Africa: Ghana's Experience from colonial times to 1998.* Praeger.

Korbéogo, G. (2018). Ordering urban agriculture: Farmers, experts, the state and the collective management of resources in Ouagadougou, Burkina Faso. *Environment and Urbanization, 30*(1), 283–300.

Kuusaana, E. D. (2017). Dynamics of winners and losers in large-scale land transactions in Ghana: Opportunities for win-win outcomes. *African Review of Economics and Finance, 9*(1), 62–95.

Kwoyiga, L. (2019). Institutional analysis of groundwater irrigation in Northeast Ghana. *African Review of Economics and Finance, 11*(2), 389–419.

Laguna, N. M. (2004). Oil policies and privatization strategies in Mexico: implications for the petrochemical sector and its production spaces. *Energy Policy, 32*, 2035–2047.

Langton, M. (2003). The "wild" the market, and the native: Indigenous people face new forms of global colonization. In W. S. Adam & M. Mulligan (Eds.), *Decolonizing nature* (pp. 79–107). Earthscan.

Langton, M. (2010). The resource curse. *Griffith Review, 28*(Autumn), 47–63. https://www.researchgate.net/publication/233990385_Langton_M_The_Resource_Curse_Griffith_Review_ed28

Laube, W. (2005). *Promise and perils of water reform: Perspectives from Northern Ghana.* ZEF Working Paper Series 10. University of Bonn.

Lawanson, T., & Oduwaye, L. (2014). Socio-economic adaptation strategies of the urban poor in the Lagos metropolis, Nigeria. *African Review of Economics and Finance, 6*(1), 139–160.

Lawrie, M., Tonts, M., & Plummer, P. (2011), Boomtowns, resource dependence and socio-economic well-being. *Australian Geographer, 42*(2), 139–164.

Lea, D. R. (1994). Lockean property rights, Tully's community ownership, and Melanesian customary communal ownership. *Journal of Social Philosophy, 25*(1), 117–132.

Lee, R. E. (Ed.). (2012). *The longue duree and world-systems analysis.* SUNY Press.

Lenger, A. (2019). The rejection of qualitative research methods in economics. *Journal of Economic Issues, 53*(4), 946–965.

Lentz, C. (2013). *Land, mobility. and belonging in West Africa.* Indiana University Press.

Lerner, J. (2009). The empirical impact of intellectual property rights on innovation: Puzzles and clues. *American Economic Review, 99*(2), 343–48.

Li, T. M. (2014). What is land? Assembling a resource for global investment. *Transactions of the Institute of British Geographers, 39*(4), 589–602.

Libecap, G. D. (2018). Policy note: Water markets as adaptation to climate change in the western United States. *Water Economics and Policy, 4*(3), 1871003. https://doi.org/10.1142/S2382624X18710030

Libecap, G. D., & Smith, J. (2001). Regulatory remedies to the common pool: The limits to oil field unitization. *Energy Journal, 22*(1), 1–26.

Libecap, G. D., & Wiggins, S. N. (1984). Contractual responses to the common pool: Prorationing of crude oil production. *The American Economic Review, 74*(1), 87–98.

Liberti, S. (2013). *Land grabbing: Journeys in the new colonialism.* Verso.

Linebaugh, P. (2008). *The Magna Carta manifesto: Liberties and commons for all.* University of California Press.

Linklater, A. (2013). *Owning the earth: The transforming history of land ownership.* Bloomsbury.

Lipton, M. (1977). *Why poor people stay poor: Urban bias in world development,* Harvard University Press.

Lu, C., & Liu, Y. (2016). Effects of China's urban form on urban air quality. *Urban Studies, 53*(12), 2607–2623.

Luong, P. J., & Weinthal, E. (2010). *Oil is not a curse: Ownership structure and institutions in Soviet successor states.* Cambridge University Press.

Lüthje, B., & Butollo, F. (2017). Why the Foxconn model does not die: Production networks and labour relations in the IT industry in South China. In A. Bieler & L. Chun-Yi (Eds.), *Chinese labour in the global economy: Capitalist exploitation and strategies of resistance* (pp. 38–53). Routledge.

Maathai, W. (2004). [Nobel Prize lecture]. The Nobel Prize. http://www .nobelprize.org/nobel_prizes/peace/laureates/2004/maathai-lecture-text.html

Maathai, W. (2011), Challenge for Africa. *Sustainability Science, 1,* 1–2.

Mabogunje, A. (1990). Urban planning and the post-colonial state in Africa: A research overview. *African Studies Review, 33*(2), 121–203.

MacDowell, L. S. (Ed.). (2017). *Nuclear portraits: Communities, the environment, and public policy.* University of Toronto Press.

MacLeavy, J., & Manley, D. (2018). (Re)discovering the lost middle: intergenerational inheritances and economic inequality in urban and regional research. *Regional Studies, 52*(10), 1435–1446.

Mahali, A., Lynch, I., Fadiji, A. W., Tolla, T., Khumalo, S., & Naicker, S. (2018). Networks of well-being in the Global South: A critical review of current scholarship. *Journal of Developing Societies, 34*(3), 1–28.

Malik, H. (2019). *Benefitting from the boom? Experiences of female business operators in resource towns of southern Queensland* [Unpublished doctoral dissertation]. University of Queensland Sustainable Minerals Institute.

Mallaby, S. (2010, July/August). The politically incorrect guide to ending poverty. *The Atlantic,* 93–104.

Manirakiza, V. (2014). Promoting inclusive approaches to address urbanisation challenges in Kigali. *African Review of Economics and Finance, 6*(1), 161–180.

Manji, A. (2013). *The politics of land reform in Africa.* Zed Books.

Manji, A. (2017). Property, conservation, and enclosure in Karura Forest, Nairobi. *African Affairs, 116*(463), 186–205.

Manning, B. R. M. (2018). *Upstream: Trust lands and power on the Feather River.* University of Arizona Press.

Manzi, T., & Smith-Bowers, B. (2005). Gated communities as club goods: Segregation or social cohesion? *Housing Studies, 20*(2), 345–359

Marais, L., Burger, P., & van Rooyen, D. (Eds.). (2018). *Mining and community in South Africa: From small town to iron town.* Routledge.

Markets in an age of anxiety. (2019, August 17). *The Economist.*

Markusen, A. (2003a). Fuzzy concepts, scanty evidence, policy distance: The case for rigour and policy relevance in critical regional studies. *Regional Studies, 37*(6 & 7), 701–717.

Markusen, A. (2003b). On conceptualization, evidence and impact: A response to Hudson, Lagendijk and Peck. *Regional Studies, 37*(6& 7), 747–751.

Marquez, A. (1990). Review of "The other path by Hernando de Soto."*Boston College Third World Law Journal, 10*(1), 204–213.

Marx, K. (1990). *Capital: Vol. 1.* Penguin. (Original work published 1867)

Marx, K., & Engels, F. (1888). *The Communist Manifesto: a road map to history's most important political document* (P. Gasper, Ed.). Haymarket Books.

Maseland, R. (2018). Is colonialism history? The declining impact of colonial legacies on African institutional and economic development. *Journal of Institutional Economics, 14*(2), 259–287.

Matsebula, V., & Yu, D. (2020). An analysis of financial inclusion in South Africa. *African Review of Economics and Finance, 12*(1), 171–202.

McFarlane, C., & Söderström, O. (2017). On alternative smart cities. *City: Analysis of Urban Trends, Culture, Theory, Policy, Action, 21*(3–4), 312–328.

McNeil, B. (2007). The costs of introducing nuclear power to Australia. *Journal of Australian Political Economy, 59*, 5–29.

Ménard, C., & Shirley, M. M. (2014). The future of new institutional economics: from early intuitions to a new paradigm? *Journal of Institutional Economics, 10*(4), 541–565

Mertzman, S. (2018, August 16). *What are rare earths, crucial elements in modern technology? 4 questions answered.* The Conversation. https://theconversation.com/what-are-rare-earths-crucial-elements-in-modern-technology-4-questions-answered-101364

Metcalfe, S., & Kepe, T. (2008). Dealing land in the midst of poverty: Commercial access to communal land in Zambia. *African and Asian Studies, 7*(2), 235–257.

Mills, S., & Sweeney, B. (2013). Employment relations in the Neostaples resource economy: Impact benefit agreements and Aboriginal governance in Canada's nickel mining industry. *Studies in Political Economy, 91*(Spring), 7–33.

Milonakis, D., & Meramveliotakis, G. (2013). Homo economicus and the economics of property rights: History in reverse order. *Review of Radical Political Economics, 45*(1), 5–23.

Ministere de L'Envionnement et Du Cadre de Vie. (2001). *Strategie et programme national de gestion durable des dechets solides.* .

Ministry of Lands and Forestry. (1999). *National land policy.*

Minsky, H. P. (1992). *The financial instability hypothesis.* Working Paper No. 74. The Jerome Levy Economics Institute of Bard College.

Mirowski, P. (1988a). *More heat than light: Economics as social physics, physics as nature's economics.* Cambridge University Press.

Mirowski, P. (1988b). Energy and energetics in economic theory: A review essay. *Journal of Economic Issues, 22*(3), 811–830.

Mitchell, A. (2015). Thinking without the "circle": Marine plastics and global ethics. *Political Geography, 47,* 77–85.

Mohee, R., & Simelane, T. (2015). *Future directions of municipal solid waste management in Africa.* Africa Institute of South Africa.

Molotch, H. (1976). The city as a growth machine. *American Journal of Sociology, 82*(2), 309–332.

Moore, S. F. (1986). *Social facts and fabrications "customary" law on Kilimanjaro, 1880–1980.* Cambridge University Press.

Morck, R., & Yeung, B. (2011). Economics, history, and causation. *Business History Review, 85*(Spring), 39–63.

Morgan, T. (2018). The techno-finance fix: A critical analysis of international and regional environmental policy documents and their implications for planning. *Progress in Planning, 119,* 1–29

Moser, S. (2020, February 21). New cities: Engineering social exclusions. *One Earth,* 125–127.

Motengwe, C., & Alagidede, P. (2017). The nexus between coal consumption, CO_2 emissions and economic growth in South Africa. *Geography Research Forum, 37,* 80–110.

Moyo, K., & Liebenberg, S. (2015). The privatization of water services: The quest for enhanced human rights accountability. *Human Rights Quarterly, 37*(3), 691–727.

Muchadenyika, D. (2020). *Seeking urban transformation: Alternative urban futures in Zimbabwe.* Weaver Press.

Munck, R., Asingwire, N., Fagan, H., & Kabonesa, C. (Eds.). (2015). *Water and development: Good governance after neoliberalism.* Zed Books.

Mundoli, S., Unnikrishnan, H., & Nagendra, H. (2019). Urban commons of the Global South: Using multiple frames to illuminate complexity. In B. Hudson, J. Rosenbloom, & D. Cole (Eds.), *Routledge handbook of the study of the commons* (pp. 220–234). Routledge.

Munro, D. (2013). Land and capital. *Journal of Australian Political Economy, 70*(Summer), 214–232.

Muralidharan, S., & Sheehan, K. (2016). "Tax" and "fee" message frames as inhibitors of plastic bag usage among shoppers: A social markting application of the theory of planned behaviour. *Social Marketing Quarterly, 22*(3), 200–217.

Murphy, J. T., & Carmody, P. (2015). *Africa's information revolution: Technical regimes and production networks in South Africa and Tanzania.* Wiley Blackwell.

Murphy, J. T., & Carmody, P. R. (2019). Generative urbanization in Africa? A sociotechnical systems view of Tanzania's urban transition. *Urban Geography, 40*(1), 128–157. https://doi.org/10.1080/02723638. (2018).1500249

Murphy, M. (2011). Africa's nuclear power potential: Its rise, recession, opportunities and constraints. In T. Simelane & M. Abdel-Rahman (Eds.), *Energy Transition in Africa* (pp. 21–54). Africa Institute of South Africa.

Mutopo, P., & Chiweshe, M. K. (2014). Water resources and biofuel production after the fast-track land reform in Zimbabwe. *African Identities, 12*(1), 124–138.

Myrdal, G. (1944). *The American dilemma: The negro problem and modern democracy.* Harper and Brothers Publishers.

Nagendra, H. (2018). The Global South is rich in sustainability lessons. *Nature, 557*(May), 485–557.

Nagendra, H. (2019). *Nature in the city: Bengaluru in the past, present, and future.* Oxford University Press.

Ndiaye, M. F. (2020). Persistent inequality in Guinea-Bissau: The role of France, the CFA Franc, and long-term currency imperialism. *African Review of Economics and Finance, 12*(1), 123–151.

Nelson, R. H. (2004). Environmental religion: Theological critique. *Case Western Reserve Law Review, 55*(1), 51–80.

Nelson, R. H. (2019). Economic religion and the worship of progress. *American Journal of Economics and Sociology, 78*(2), 319–362.

Newell, S. (2012). *The modernity bluff: Crime, consumption, and citizenship in Côte d'Ivoire.* University of Chicago Press.

Newman, C., Page, J., Rand, J., Shimeles, A., Söderbom, M., & Tarp, F. (2016). *Manufacturing transformation: Comparative studies of industrial development in Africa and emerging Asia.* Oxford University Press.

Newman, D. (2015). The car and the commons. *Review of Radical Political Economics, 48*(1), 53–65.

Nibi, S. H. (2012). *A feminist political ecology of large-scale agrofuel production in Northern Ghana: A case study of Kpachaa* [Master's thesis, International Institute of Social Studies]. Erasmus University Thesis Repository. http://hdl.handle.net/2105/13147

Niman, N. B. (2011). Henry George and the intellectual foundations of the open source movement. *American Journal of Economics and Sociology, 70*(4), 904–927.

Njeru, J. (2006). The urban political ecology of plastic bag waste problem in Nairobi, Kenya. *Geoforum, 37,* 1046–1058.

Njoh, A. J. (2009). Ideology and public health elements of human settlement policies in Sub-Saharan Africa. *Cities, 26,* 9–18.

Njoh, A. J. (2014). *Urban planning and public health in Africa: Historical, theoretical and practical dimensions of a continent's water and sanitation problematic.* Ashgate.

Njoh, A. J. (2016). *French urbanism in foreign lands.* Springer.

Njoh, A. J., & Chie, E. P. (2019). Vocabularies of spatiality in French colonial urbanism: Some covert rationales of street names in colonial Dakar, West

Africa and Saigon, Indochina. *Journal of Asian and African Studies, 54*(8), 1109–1127.

Noorloos, V. F., Klaufus, C., & Steel, G. (2019). Land in urban debates: Unpacking the grab-development dichotomy. *Urban Studies, 56*(5), 855–867.

Norberg, J. (2005). *In defence of global capitalism*. The Center for Independent Studies.

North, D. C. (1991). Institutions. *The Journal of Economic Perspectives, 5*(1), 97–112.

Nukunya, G. K. (2003). *Tradition and change in Ghana: An introduction to sociology* (2nd ed.). Ghana Universities Press.

Nwoke, C. N. (1984). *The global struggle over surplus profit for mining: A critical extension of Marx's rent theory* [Unpublished doctoral dissertation]. Graduate School of International Studies, University of Denver.

Nyamnjoh, F. (2019). Decolonizing the university in Africa. *Oxford Research Encyclopedia of Politics*. https://doi.org/10.1093/acrefore/9780190228637.013.717

Nyamnjoh, F. B. (2012). "Potted plants in greenhouses": A critical reflection on the resilience of colonial education in Africa. *Journal of Asian and African Studies, 47*(2), 129–154.

Nyamnjoh, F. B. (2015). *C'est l'homme qui fait l'homme: Cul-de-sac Ubuntu-ism in Côte d'Ivoire*. Langaa Research & Publishing CIG.

Nyantakyi-Frimpong, H. (2020). What lies beneath: Climate change, land expropriation, and zaï agroecological innovations by smallholder farmers in Northern Ghana. *Land Use Policy, 92*, 1–11. https://doi.org/10.1016/j.landusepol.2020.104469

Nyeleni Conference. (2011, November 23). *Conference declaration: Stop land-grabbing now!* La Via Campesina. https://viacampesina.org/en/stop-land-grabbing-now/

Obeng-Odoom, F. (2010a). An urban twist to politics in Ghana. *Habitat International, 34*(4), 392–399.

Obeng-Odoom, F. (2010b). Avoid the oil curse in Ghana: Is transparency sufficient? *African Journal of International Affairs, 13*(1&2), 89–119.

Obeng-Odoom, F. (2011a). Developing Accra for all? The story behind Africa's largest millennium city. *Development: A Journal of the Society for International Development, 54*(3), 384–392.

Obeng-Odoom, F. (2011b). Real estate agents in Ghana: A suitable case for regulation? *Regional Studies, 45*(3), 403–416.

Obeng-Odoom, F. (2011c). The informal sector in Ghana under siege. *Journal of Developing Societies, 27*(3&4), 355–392.

Obeng-Odoom, F. (2012a). Review of "Envisioning real utopias." *Capital and Class, 36*(1), 184–187.

Obeng-Odoom, F. (2012b). Land reform in Africa: Theory, practice, and outcome. *Habitat International, 36*(1), 161–170.

Obeng-Odoom, F. (2013a). The state of African cities 2010: Governance, inequality, and urban land markets. *Cities, 31*(April), 425–429.

Obeng-Odoom, F. (2013b). *Governance for pro-poor urban development: Lessons from Ghana*. Routledge.

Obeng-Odoom, F. (2013c). The mystery of capital or the mystification of capital? *Review of Social Economy, 71*(4), 427–442.

Obeng-Odoom, F. (2013d). Do African cities create markets for plastics or plastics for markets? *Review of African Political Economy, 40*(137), 466–474.

Obeng-Odoom, F. (2013e). Managing land for the common good? Evidence from a community development project at Agona, Ghana. *Journal of Pro-Poor Growth, 1*(1), 29–46.

Obeng-Odoom, F. (2013f). The grab of the world's land and water resources. *Revista de Economia Politica, 33*(3), 527–537.

Obeng-Odoom, F. (2014a). Measuring what? "Success" and "failure" in Ghana's oil industry. *Society and Natural Resources, 27*(6), 656–670.

Obeng-Odoom, F. (2014b). *Oiling the urban economy: Land, labour, capital and the state in Sekondi-Takoradi, Ghana*. Routledge.

Obeng-Odoom, F. (2014c). Urban land policies in Ghana: A case of the emperor's new clothes? *The Review of Black Political Economy, 41*(2), 119–143.

Obeng-Odoom, F. (2014d). Green neoliberalism: Recycling and sustainable urban development in Sekondi-Takoradi. *Habitat International, 41*, 129–134.

Obeng-Odoom, F. (2015a). Understanding land grabs in Africa: Insights from Marxist and Georgist political economics. *The Review of Black Political Economy, 42*(4), 337–354.

Obeng-Odoom, F. (2015b). A little fuel for an African-Australian relationship? *Journal of Economic Issues, 49*(3), 865–871.

Obeng-Odoom, F. (2015c). Africa: On the rise but to where? *Forum for Social Economics, 44*(3), 234–250.

Obeng-Odoom, F. (2015d). Global political economy and frontier economies in Africa: Implications from the oil and gas industry in Ghana. *Energy Research and Social Science, 10*(November), 41–56.

Obeng-Odoom, F. (2015e, May). *Oil rents, policy, and social development: Lessons from the Ghana controversy*. United Nations Research Institute for Social Development Research Paper 2.

Obeng-Odoom, F. (2016a, January 7). *Why the SDGs need institutional political economy for inclusive, resilient cities*. United Nations Research Institute for Social Development Think Piece.

Obeng-Odoom, F. (2016b). *Reconstructing urban economics: Towards a political economy of the built environment*. Zed Books.

Obeng-Odoom, F. (2016c). Sustainable devlopment: A Georgist perspective. In A. Allen, A. Lampis, & M. Swillng (Eds.), *Untamed urbanism* (pp. 191–203). Routledge.

Obeng-Odoom, F. (2016d). Urban governance in Africa today: Conceptualisation, trends, and innovation. *Growth and Change, 48*(1), 4–21.

Obeng-Odoom, F. (2016e). Marketising the commons in Africa: The case of Ghana. *Review of Social Economy, 74*(4), 390–419.

Obeng-Odoom, F. (2017a). Teaching property economics students political economy: Mission impossible? *International Journal of Pluralism and Economics Education, 8*(4), 359–377.

Obeng-Odoom, F. (2017b). The wretched of the earth. *Journal of Australian Political Economy, 78*, 5–23.

Obeng-Odoom, F. (2018). Transnational corporations and urban development. *American Journal of Economics and Sociology, 47*(2), 447–510.

Obeng-Odoom, F. (2019a). Pedagogical pluralism in undergraduate urban economics education. *International Review of Economics Education, 31*, Article 100158. https://doi.org/10.1016/j.iree.2019.100158

Obeng-Odoom, F. (2019b). Economics, education, and citizenship. *Australian Universities' Review, 61*(1), 3–11.

Obeng-Odoom, F. (2019c). Petroleum accidents in the Global South. *Research in Political Economy, 33*, 111–142.

Obeng-Odoom, F. (2019d). Economic cycles, economic crises, resource grabs, and expulsion. *International Critical Thought, 9*(1), 64–84.

Obeng-Odoom, F. (2020). *Property, Institutions, and Social Stratification in Africa.* Cambridge University Press.

Obeng-Odoom, F. (2021). From commons to gifts. Unpublished Manuscript.

Obeng-Odoom, F., & Bromley, D. W. (2020). Interview with Professor Daniel W. Bromley. *African Review of Economics and Finance, 12*(1), 38–60.

Obeng-Odoom, F., & Gyampo, R. E. V. (2017). Land grabbing, land rights, and the role of the courts. *Geography Research Forum, 31*, 127–147.

O'Boyle, E. J., & Welch, P. J. (2016). Tracing the origins of personalist economics to Aristotle and Aquinas. *Forum for Social Economics, 45*(1), 3–18.

O'Brien, D. T. (2018). The urban commons: How data and technology can rebuild our communities. Harvard University Press.

Obrist, B., Cissé, G., Koné, B., Dongo, K., Granado, S., & Tanner, M. (2006). Interconnected slums: Water, sanitation and health in Abidjan, Côte d'Ivoire. *The European Journal of Development Research, 18*(2), 319–336.

O'Connor, J. (1988). Capitalism, nature, socialism a theoretical introduction. *Capitalism Nature Socialism, 1*(1), 11–38.

O'Connor, J. (1991). On the two contradictions of capitalism. *Capitalism Nature Socialism, 2*(3), 107–109.

Odendaal, N. (2012). Reality check: Planning education in the African urban century. *Cities, 29*(3), 174–182.

O'Donnell, E. T. (2015). *Henry George and the crisis of inequality: Progress and poverty in the gilded age.* Columbia University Press.

Offer, A., & Söderberg, G. (2016). *The Nobel factor: The Prize in Economics, social democracy, and the market turn.* Princeton University Press.

Okyerea, D. K., Poku-Boansi, M., Adarkwa, K. K. (2018). Connecting the dots: The nexus between transport and telecommunication in Ghana. *Telecommunications Policy, 42*(10), 836–844.

Ojong, N. (2011). Livelihood strategies in African cities: The case of residents in Bamenda, Cameroon. *African Review of Economics and Finance, 3*(1), 8–24.

Ojong, N. (2020). *The everyday life of the poor in Cameroon: The role of social networks in meeting needs.* Routledge.

Okoth-Ogendo, H. W. O. (2003). The tragic African commons: A century of expropriation, suppression and subversion. *University of Nairobi Law Journal,* 1(1), 107–117.

Olaniyi, R. (2008). Review of "The pan-African nation: Oil and the spectacle of culture in Nigeria." *African Studies Quarterly, 10*(2 & 3), 198–200. https://sites.clas.ufl.edu/africanquarterly/files/Books-Reviews-Vol10Issue23.pdf#page=46.

Ollenu, N. A. (1971). The changing law and law reform in Ghana. *Journal of African Law, 15*(2), 132–181.

Olukoju, A. (2004). "Never expect power always": Electricity consumers' response to monopoly, corruption and inefficient services in Nigeria. *African Affairs, 103,* 51–71.

Oquaye, M. (1995a). The GhanaianeElections of 1992 – A dissenting view. *African Affairs, 94*(375), 259–275.

Oquaye, M. (1995b). Human rights and the transition to democracy under the PNDC in Ghana. *Human Rights Quarterly, 17*(3), 556–573.

Orwin, C. S. (1938). Observations on the open fields. *The Economic History Review, 8*(2), 125–135.

Orwin, C. S., & Orwin, C. S. (1967). *The Open Fields,* Clarendon Press.

Ostrom, E. (1990). *Governing the commons: The evolution of institutions for collective action.* Cambridge University Press.

Ostrom, E. (2007). Challenges and growth: the development of the interdisciplinary field of institutional analysis. *Journal of Institutional Economics, 3*(3), 239–264.

Ostrom, E. (2008). Institutions and the environment. *Economic Affairs, 28*(3), 24–31.

Ostrom, E. (2009a, December 8). *Beyond markets and states: Polycentric governance of complex economic systems* [Nobel Prize lecture]. The Nobel Prize. https://www.nobelprize.org/prizes/economic-sciences/2009/ostrom/lecture/

Ostrom, E. (2009b). *A polycentric approach for coping with climate change.* Background paper to the 2010 World Development Report, Policy Research Working Paper 5095. World Bank.

Ostrom, E. (2010a). Beyond markets and states: Polycentric governance of complex economic systems. *American Economic Review,* (June), 641–672.

Ostrom, E. (2010b). Polycentric systems for coping with collective action and global environmental change. *Global Environmental Change, 20,* 550–557.

Ostrom, E. (2010c). A long polycentric journey. *Annual Review of Political Science, 13*(1), 1–23.

Ostrom, E. (2010d). Organizational economics: Applications to metropolitan governance. *Journal of Institutional Economics, 6*(1), 109–115.

Ostrom, E. (2012a). *Green from the grassroots.* Project Syndicate. http://www .project-syndicate.org/commentary/green-from-the-grassroots

Ostrom, E. (2012b). *The future of the commons: Beyond market failure and government regulation.* The Institute of Economics Affairs.

Ostrom, E., & Basurto, X. (2011). Crafting analytical tools to study institutional change. *Journal of Institutional Economics, 7*(3), 317–343

Ostrom, E., Walker, J., & Gardner, R. (1992). Covenants with and without a sword: Self-governance is possible. *The American Political Science Review, 86*(2), 404–417.

O'Sullivan, A. (2012). *Urban economics.* McGraw-Hill Irwin.

Otiso, K. M., Derudder, B., Bassens, D., Devriendt, L., & Witlox, F. (2011) Airline connectivity as a measure of the globalization of African cities. *Applied Geography,* 31, 609–620.

Our urban future. (2020). [Editorial]. *One Earth,* 2, 111–112. https://doi.org/ 10.1016/j.oneear.2020.02.006

Overå, R. (2001). *Institutions, mobility and resilience in the Fante migratory fisheries of West Africa.* CMI Working Paper 2001:2, Chr. Michelsen Institute. https://www .cmi.no/publications/900-institutions-mobility-and-resilience-in-the-fante

Owusu-Ofori, B., & Obeng-Odoom, F. (2015). The ravages of resettlement: A Ghanaian case study. *Social Change, 45*(2), 234–241.

Oyake-Ombis, L. (2012). *Managing plastic waste in urban Kenya: Niche innovations in production and recycling* [Doctoral dissertation, Wageningen University]. Wageningen UR E-depot. https://edepot.wur.nl/239452

Paller, J. (2019). *Democracy in Ghana: Everyday politics in urban Africa.* Cambridge University Press.

Palmer, M. (2017, October). Good uses. *Financial Times,* 5.

Papadimitropoulos, V. (2018). Reflections on the contradictions of the commons. *Review of Radical Political Economics, 50*(2), 317–331.

Pearce, A., & Stilwell, F. (2008). "Green collar" jobs: Employment impacts of climate change policies. *Journal of Australian Political Economy, 62,* 120–138.

Pearce, F. (2012). *The land grabbers: The new fight over who owns the earth.* Beacon Press.

Pearson, N., & Kostakidis-Lianos, L. (2004). *Building Indigenous capital: removing obstacles to participation in the real economy.* Position Paper, Cape York Institute for Policy and Leadership.

Peck, J. (2015). Chicago-school suburbanism. In P. Hamel & R. Keil (Eds.), *Suburban governance: A global view* (pp. 130–152). University of Toronto Press.

Peirce, W. (2015). Henry George and Jane Jacobs on the sources of economic growth. *American Review of Economics and Sociology, 74*(3), 510–530.

Pejovich, S. (1972). Towards an economic theory of the creation and specification of property rights. *Review of Social Economy, 30*(3), 309–325.

Pennington, M. (2012). Elinor Ostrom, common-pool resources and the classical liberal tradition. In *The future of the commons: Beyond market failure and government regulation* (pp. 21–47). The Institute of Economics Affairs.

Pennington, M. (2013). Elinor Ostrom and the robust political economy of common-pool resources. *Journal of Institutional Economics, 9*(4), 449–468.

Peters, S. (2017). Beyond curse and blessing: Rentier society in Venezuela. In B. Engels & K. Dietz (Eds.), *Contested extractivism, society, and the state: Struggles over mining and land* (pp. 45–68). Palgrave.

Petrella, F. (1981). Henry George, the classical model and technological change: The ignored alternative to the single tax in *Progress and Poverty*. *American Journal of Economics and Sociology, 40*(2), 191–206.

Pickbourn, L. (2020). When women go to the city: African women's rural-urban migration and the Sustainable Development Goals. In M. Konte & N. Tirivayi (Eds.), *Women and sustainable human development. gender, development and social change*. Palgrave Macmillan.

Pithouse, R. (2014). An urban commons? Notes from South Africa. *Community Development Journal, 49*(Suppl. 1), i31–i43.

Polanyi, K. (1957). The economy as instituted process. In K. Polanyi, C. M. Arensberg, & H. W. Pearson (Eds.), *Trade and market in the early empires* (pp. 243–270). Free Press.

Polanyi, K. (2001). *The great transformation: the political and economic origins of our time*. Beacon Press. (Original work published 1944)

Porter, L. (2011). Informality, the commons and the paradoxes for planning: Concepts and debates for informality and planning. *Planning Theory and Practice, 12*(1), 115–153.

Potts, J. (2018). Governing the innovation commons. *Journal of Institutional Economics, 14*(6), 1025–1047.

Potts, J., & Hartley, J. (2015). How the social economy produces innovation. *Review of Social Economy, 73*(3), 263–282.

Prato, S. (2014). Shared societies: A new approach to planetary coniventia [Editorial]. *Development, 57*(1), 1–7.

Pratt, M. L. (1990). Fieldwork in common places. In J. Clifford & G. E. Marcus (Eds.), *Writing culture: The poetics and politics of ethnography* (pp. 27–50). Oxford University Press.

Pullen, J. (2013). An essay on distributive justice and the equal ownership of natural resources. *American Journal of Economics and Sociology, 72*(5), 1044–1074.

Pullen, J. (2014). *Nature's gifts: The Australian lectures of Henry George on the ownership of land and other natural resources*. Desert Pea Press.

Pullen, J. (2019). The Pope and Henry George: Pope Leo XIII compared with Henry George on the ownership of land and other natural resources. A possible rapprochement? *Solidarity, 9*(1). https://researchonline.nd.edu.au/cgi/viewcontent.cgi?article=1116&context=solidarity

Ram, A. (2017, October). Boot camp rebels. *Financial Times,* 6–7.

Razif, N. F. M., Ab Halim, A., Samsulkamal, N. S., & Wahab, N. A. (2017). Real estate market in Islamic history to modern era: The origin and evolution of housing speculation. *Journal of al-Tamaddun, 12*(2), 127–137. https://doi.org/10.22452/JAT.vol12no2.10

Redford, A. (2020). Property rights, entrepreneurship, and economic development. *The Review of Austrian Economics, 33,* 139–161. https://doi.org/10.1007/s11138-019-00485-6

Renom, J. G., Mwamidi, D. M., & Domínguez, P. (2020). Holistic ethnographies of East African customary pastoral commons needed? *Current Opinion in Environmental Sustainability, 43,* 83–90.

Research & Degrowth. (2010). Degrowth Declaration of the Paris 2008 conference. *Journal of Cleaner Production, 18*(6), 523–524.

Rhodes, R. (2018). *The making of the atomic bomb: Energy, a human history*. Simon and Schuster.

Richardson, T., & Weszkalnys, G. (2014). Introduction: Resource materialities. *Anthropological Quarterly, 87*(1), 5–30.

Robbins, G. (2012). Mining FDI and urban economies in Sub-Saharan Africa: Exploring the possible linkages. *Local Economy, 28*(2), 158–169.

Robert-Nicoud, F. (2008). Offshoring of routine tasks and (de)industrialization: Threat or opportunity – And for whom? *Journal of Urban Economics, 63*(2), 517–535.

Robertson, C. (1983). The death of Makola and other tragedies. *Canadian Journal of African Studies, 17*(3), 469–495.

Robinson, J. (2006). *Ordinary cities: Between modernity and development*. Routledge.

Robra, B., & Heikkurinen, P. (2019). Degrowth and the sustainable development goals. In W. Leal Filho, A. M. Azul, L. Brandli, P. G. Özuyar, & T. Wall (Eds.), *Decent work and economic growth: Encyclopedia of the UN Sustainable Development Goals*. Springer Nature.

Rodney, W. (2011). *How Europe underdeveloped Africa*. Black Classic Press. (Original work published 1972)

Rodríguez-Labajos, B., Yánez, I., Bond, P., Grey, L., Munguti, S., Ojo, G. U., Overbeek, W. (2019). Not so natural an alliance? Degrowth and environmental justice movements in the Global South. *Ecological Economics, 157,* 175–184.

Romer, P. (2010). Technologies, Rules, and Progress: The Case for Charter Cities. Centre for Global Development, www.cgdev.org/content/publications/detail/1423916 (accessed 22.10.2018).

Romer, P. M. (2018, December 8). *On the possibility of progress* [Nobel Prize lecture]. The Nobel Prize. https://www.nobelprize.org/prizes/economic-sciences/2018/romer/lecture/

Rose, C. M. (1986). *The comedy of the commons: commerce, custom, and inherently public property.* Faculty Scholarship Series Paper 1828. http://digitalcommons.law.yale.edu/fss_papers/1828

Rosewarne, S. (2002). Towards an ecological political economy. *Journal of Australian Political Economy, 50*(December), 179–199.

Rosewarne, S. (2011). Meeting the challenge of climate change: The poverty of the dominant economic narrative and market solutions as subterfuge. *Journal of Australian Political Economy, 66,* 17–50.

Ross, C. (2017). *Ecology and power in the age of empire: Europe and the transformation of the tropical world.* Oxford University Press.

Ryan-Collins, J., Lloyd, T., & MacFarlane, L. (2017). *Rethinking the economics of land and housing.* Zed Books.

Sachs, W. (Ed.). (2010). *The development dictionary.* Zed Books.

Sackeyfio-Lenoch, N. (2014). *The politics of chieftaincy: Authority and property in colonial Ghana, 1920–1950.* University of Rochester Press.

Sanni, M., Oladipo, O. G., Ogundari, I. O., & Aladesanni, O. T. (2014). 'Adopting latecomers' strategies for the development of renewable energy technology in Africa. *African Journal of Science, Technology, Innovation and Development, 6*(4), 253–263.

Sarker, A., & Blomquist, W. (2019). Addressing misperceptions of *Governing the Commons. Journal of Institutional Economics, 15*(2), 281–301. https://doi.org/10.1017/S1744137418000103

Schaefer, B. D. (2007, March 23). *The crisis in Zimbabwe: How the U. S. should respond.* The Heritage Foundation. http://www.heritage.org/research/reports/2007/03/the-crisis-in-zimbabwe-how-the-us-should-respond#_ftn43

Schenck, R., & Blaauw, P. F. (2011). The work and lives of street waste pickers in Pretoria – A case study of recycling in South Africa's urban informal economy. *Urban Forum, 22,* 411–430.

Schläppi, D. (2016, May 11). *Shared ownership as key issue of Swiss history: Common pool resources, common property institutions and their impact on the political culture of Switzerland from the beginnings to our days* [Keynote address]. IASC-European Regional Conference, Bern, Switzerland.

Schläppi, D. (2019). Shared ownership as a key issue of Swiss history: Common-pool resources, common property institutions and their impact on the political culture of Switzerland from the beginnings to our days. In T. Haller, T. Breu, T. de Moor, C. Rohr, & H. Znoj (Eds.), *The commons in a global world: Global connections and local responses* (pp. 23–33). Routledge.

Schlatter, R. (1951). *Private property: The history of an idea.* Russell and Russell.

Schoneveld, G., German, L., & Nutakor, E. (2011). Land-based investments for rural development? A grounded analysis of the local impacts of biofuel feedstock plantations in Ghana. *Ecology and Society, 16*(4), 10. https://doi .org/10.5751/ES-04424-160410.

Schultz, T. W. (1951), A framework for land economics: The long view. *American Journal of Agricultural Economics, 33*(2), 204–215.

Sebastian, A. G., & Warner, J. F. (2014). Geopolitical drivers of foreign investment in African land and water resources. *African Identities, 12*(1), 8–25.

Seccareccia, M., & Correa, E. (2017). Supra-national money and the Euro crisis: Lessons from Karl Polanyi. *Forum for Social Economics, 46*(3), 221–320.

Segal, P. (2011). Resource rents, redistribution, and halving global poverty: The resource dividend. *World Development, 39*(4), 475–489.

Semuels, A. (2019, March 5). Is this the end of recycling? *The Atlantic.* https:// www.theatlantic.com/technology/archive/2019/03/china-has-stopped- accepting-our-trash/584131/

Sen, A. (2015). *Development research and changing priorities* [UNU-WIDER Annual Lecture 19]. UNU-WIDER. https://www.wider.unu.edu/sites/ default/files/AL19-2015.pdf

Sharma, D. (2012). *A new institutional economics approach to water resource management* [Docotral dissertation, University of Sydney]. Semantics Scholar. https://pdfs.semanticscholar.org/e084/74bedfa04055398164ef5534 949179a14e81.pdf

Sherman, H. J. (1993). The historical approach to political economy. *Review of Social Economy, 51*(3), 302–322.

Shinwell, M., & Cohen, G. (2020). Measuring countries' progress on the Sustainable Development Goals: Methodology and challenges. *Evolutionary and Institutional Economics Review, 17*(1), 167–182.

Shipton, P. (2007). *The nature of entrustment: Intimacy, exchange, and the sacred in Africa.* Yale University Press.

Shipton, P. (2009). *Mortgaging the ancestors: Ideologies of attachment in Africa.* Yale University Press.

Shipton, P. (2010). *Credit between cultures: Farmers, financiers, and misunderstanding in Africa.* Yale University Press.

Showers, K. B. (2014). Europe's long history of extracting African renewable energy: Contexts for African scientists, technologists, innovators and policy-makers. *African Journal of Science, Technology, Innovation and Development, 6*(4), 301–313.

Showers, K. B. (2019). Biofuels' unbalanced equations: Misleading statistics, networked knowledge and measured parameters. Part I: Evolution of globalised soil, land and terrain databases. *International Review of Environmental History, 5*(1), 61–83.

Shrader-Frechette, K. (2011). *What will work: Fighting climate change with renewable energy, not nuclear power.* Oxford University Press.

Shrubsole, G. (2019). *Who owns England? How we lost our green and pleasant land, and how to take it back.* William Collins.

Siba, E., & Sow, M. (2017, November 1). Smart city initiatives in Africa. *Brookings: The Brookings Institution Official Blog.* https://www.brookings.edu/blog/africa-in-focus/2017/11/01/smart-city-initiatives-in-africa/

Simelane, T., & Abdel-Rahman, M. (Eds.). (2011). *Energy transition in Africa.* Africa Institute of South Africa.

Sjaastad, E., & Cousins, B. (2008). Formalisation of land rights in the South: An overview. *Land Use Policy, 26,* 1–9.

Small, G. (2004). Property, commerce, and living God's will. *Journal of Interdisciplinary Studies, 16*(1 & 2), 157–172.

Small, G. (2013). Property, economics, and God. *Culture Wars, 32,* 7.

Smith, A. (2005). *The theory of moral sentiments* (6th ed.). https://www.ibiblio.org/ml/libri/s/SmithA_MoralSentiments_p.pdf (Original work published 1790)

Smith, A. (2007). *An Inquiry in the nature and causes of the wealth of nations.* Modern Library. https://www.ibiblio.org/ml/libri/s/SmithA_Wealth Nations_p.pdf (Original work published 1776)

Smith, L. T. (2012). *Decolonizing methodologies.* Zed Books; Otago University Press.

Solari, S. (2017). Roman Catholicism and the founding principles of liberalism: Liberty and private property. *Forum for Social Economics.* https://doi.org/10.1080/07360932.2017.1402358

Sorenson, S. B., Morssink, C., & Campos, P. A. (2011). Safe access to safe water in low-income countries: Water fetching in current times. *Social Science and Medicine, 72,* 1522–1526.

South African Population Research Infrastructure Network. (2002). *The policy roots of economic crisis and poverty: A multi-country participatory assessment of structural adjustment.*

Special report: Canada. (2019, July 17). *The Economist,* 3–12.

Spies-Butcher, B., & Stilwell, F. (2009). Climate change policy and economic recession. *Journal of Australian Political Economy, 83,* 108–125.

Spies-Butcher, B., Paton, J., & Cahill, D. (2012). *Market society: History, theory, practice.* Melbourne University Press.

Squires, S. (2013). *Urban and environmental economics: An introduction.* Routledge.

Stacey, P. (2019). *State of slum: Precarity and informal governance at the margins in Accra.* Zed Books.

Stacey, P., & Lund, C. (2016). In the state of slum: governance in an informal urban settlement in Ghana. *Journal of Modern African Studies, 54*(4), 591–695.

Stavrides, S. (2016). *The city as commons,* Zed Books.

Stevens, E. S. (2002). *Green plastics: An introduction to the new science of biodegradable plastics.* Princeton Univers Press.

Stevis, D., Uzzell, D., & Rathzel, N. (2018). The labour-nature relationship: varieties of labour environmentalism. *Globalizations, 15*(4), 439–453.

Stilwell, F. (1992). *Understanding cities and regions: Spatial political economy.* Pluto Press.

Stilwell, F. (1999), Cost-benefit analysis. In P. A. O'Hara (Ed.), *Encyclopedia of political economy* (pp. 157–162). Routledge.

Stilwell, F. (2000). Towards sustainable cities. *Urban Policy and Research, 18*(2), 205–218.

Stilwell, F. (2006). *Political economy: The contest of economic ideas.* Oxford University Press.

Stilwell, F. (2011). Selling the environment in order to save it? In L. Chester, M. Johnson, & P. Kriesler (Eds.), *Contemporary issues for heterodox economists.* University of New South Wales.

Stilwell, F. (2012a). *Political economy: The contest of economic ideas.* Oxford University Press.

Stilwell, F. (2012b). Marketising the environment. *Journal of Australian Political Economy, 68,* 108–127.

Stilwell, F. (2017). Why emphasize economic inequality in development? *Journal of Australian Political Economy, 78,* 24–47.

Stilwell, F. (2019a). From economics to political economy: Contradictions, challenge and change. *American Journal of Economics and Sociology, 78*(1), 35–62.

Stilwell, F. (2019b). *The political economy of inequality.* Polity Press.

Stilwell, F., & Jordan, K. (2004a). The political economy of land: Putting Henry George in his place. *Journal of Australian Political Economy, 54*(December), 119–134.

Stilwell, F., & Jordan, K. (2004b). *Land tax: A green policy priority?* Discussion paper prepared for the Greens Economic Policy Group, Sydney.

Stilwell, F., & Primrose, D. (2010). Economic stimulus and restructuring: Infrastructure, green jobs and spatial impacts. *Urban Policy and Research, 28*(1), 5–25.

Stoler, J. (2012). Improved but unsustainable: Accounting for sachet water in post-2015 goals for global safe water. *Tropical Medicine and International Health, 17*(12), 1506–1508.

Stoler, J., Tutu, R. A., & Winslow, K. (2015). Piped water flows but sachet consumption grows: The paradoxical drinking water landscape of an urban slum in Ashaiman, Ghana. *Habitat International, 47,* 52–60.

Stowers, G. N. L. (2018). *Managing the sustainable city.* Routledge.

Subere-Albawy, F. (2015). Parks and nature reserves, not vacant lots, bring nature. *Progress, 1113*(Autumn), 15–16.

Sullivan, T. (2008). *The Church of the Empire versus The Christian Church of North Africa, 312–430 A.D.* Radical Christian Press.

Sun, L., Li, H., Dong, L., Fang, K., Ren, J., Geng, Y., Fujii, M., Zhang, W., Zhang, N., & Liu, Z. (2017). Eco-benefits assessment on urban industrial symbiosis based on material flows analysis and emergy emergy evaluation approach: A case of Liuzhou city, China. *Resources, Conservation and Recycling, 119,* 78–88. https://doi.org/10.1016/j.resconrec.2016.06.007

Sundström, L. (1974). *The exchange economy of pre-colonial tropical Africa.* C. Hurst and Company.

Swyngedouw, E. (2015). Urbanization and environmental futures: Politicizing urban political ecologies. In T. Perreault, G. Bridge, & J. McCarthy (Eds.), *The Routledge handbook of political ecology* (pp. 609–619). Routledge.

Tabachnick, D. (2016). Two models of ownership: How commons has co-existed with private property. *The American Journal of Economics and Sociology, 75*(2), 488–563.

Talukdar, R. (2017). Hiding neoliberal coal behind the Indian poor. *Journal of Australian Political Economy, 78,* 132–158.

Tarko, V. (2012). Elinor Ostrom's life and work. In *The future of the commons: Beyond market failure and government regulation* (pp. 48–67). The Institute of Economics Affairs.

Tarko, V. (2017). *Elinor Ostrom: An intellectual biography.* Rowman and Littlefield.

Taxing carbon. (2020, May 23). *The Economist.*

Theesfeld, I. (2019). A structured checklist to identify connections between land grabbing and water grabbing. In T. Haller, T. Breu, T. de Moor, C. Rohr, & H. Znoj (Eds.), *The commons in a global world: Global connections and local responses* (pp. 437–453). Routledge.

Theobald, P. (1997). *Teaching the commons: Place, pride, and the renewal of community.* Westview Press.

Thirsk, J. (1964, December). The common fields. *Past and Present, 29,* 3–25.

Thynell, M. (2018). Urban inequality in a fragile global city: The case of Jakarta. In J Hellman, M. Thynell, & R. Van Voorst (Eds.), *Jakarta claiming spaces and rights in the city* (Chapter 2). Routledge.

Tiebout, C. (1956). A pure theory of local expenditures. *The Journal of Political Economy, 64*(5), 416–424.

Tignino, M. (2014). The right to water and sanitation in post-conflict legal mechanisms: An emerging regime? In E. Weinthal, J. Troell, & M. Nakayama (Eds.), *Water and post-conflict peacebuilding* (pp. 383–402). Earthscan.

Toivanen, T., & Kröger, M. (2019). The role of debt, death and dispossession in world-ecological transformations: Swidden commons and tar capitalism in nineteenth-century Finland. *The Journal of Peasant Studies, 46*(7), 1368–1388.

Tonah, S. (2005). *Fulani in Ghana: Migration history, integration and resistance.* University of Ghana; Yamens Press.

To the last drop. (2019, November 2). *The Economist, 13,* 62–62.

Traore, N. (2000, June 19–22). *Financing the urban poor: SODECI's experience in Côte d'Ivoire* [Paper presentation]. Nairobi.

Tremann, C. (2013). Temporary Chinese migration to Madagascar: Local perspectives, economic impacts, and human capital flows. *African Review of Economics and Fiannce, 5*(1), 9–20.

Tsamenyi, M. (2013). *Analysis of the adequacy of legislative framework in Ghana to support fisheries co-management and suggestions for a way forward.* Coastal Resources Center, University of Rhode Island; USAID Integrated Coastal and Fisheries Governance Program for the Western Region of Ghana.

Tsey, K., & Short, S. (1995). From head loading to the iron horse: The unequal health consequences of railway construction and expansion in the Gold Coast, 1898–1929. *Social Science and Medicine, 40*(5), 613–621.

Tsikata, D. (2006). *Living in the shadow of the large dams.* Brill.

Tsikata, D., & Yaro, J. A. (2014). When a good business model is not enough: Land transactions and gendered livelihood prospects in rural Ghana. *Feminist Economics, 20*(1), 202–226.

Turner, J. F. C. (1976). *Housing by people: Towards autonomy in building environments.* Pantheon Books.

Turpie, J. K., Marais, C., & Blignaut, J. N. (2008). The working for water programme: Evolution of a payments for ecosystem services mechanism that addresses both poverty and ecosystem service delivery in South Africa. *Ecological Economics, 65*(4), 788–798.

Tymoigne, E. (2003). Keynes and commons on money. *Journal of Economic Issues, 37*(3), 527–545.

Ubink, J., & Quan, J. (2008). How to combine tradition and modernity? Regulating customary land management in Ghana. *Land Use Policy, 25*(2), 198–213.

UN Department of Economic and Social Affairs. (2010). *World urbanisation prospects, 2009 revision.*

UN-HABITAT. (2001). *The state of the world's cities.*

UN-HABITAT. (2007). *Enhancing urban safety and security.* Earthscan.

UN-HABITAT. (2008). *State of the world's cities, 2010/2011.*

United Nations Economic Commission for Africa. (2011). *Minerals and Africa's development: The International Study Group Report on Africa's mineral regimes.* Addis Ababa. http://www.uneca.org/sites/default/files/PublicationFiles/mineral_africa_development_report_eng.pdf

United Nations Environment Programme. (2015). *Côte d'Ivoire: Post-Conflict Environmental Assessment.*

United Nations Office of the High Commissioner on Human Rights. (2016). *Draft report on the Second Session of the Open-Ended Intergovernmental Working Group on Transnational Corporations and Other Business Enterprises with Respect to Human Rights.* http://www.ohchr.org/EN/HRBodies/HRC/WGTransCorp/Session2/Pages/Session2.aspx

van Griethuysen, P. (2012). Bona diagnosis, bona curatio: How property economics clarifies the degrowth debate. *Ecological Economics, 84,* 262–269.

van Laerhoven, F., & Ostrom, E. (2007). Traditions and trends in the study of the commons. *International Journal of the Commons, 1*(1), 3–28.

Veblen, T. (2009). *Absentee ownership: Business enterprise in recent times: The case of America.* Transactions Publishers. (Original work published 1923)

Viscusi, W. K., Huber, J., & Bell, J. (2012). Alternative policies to increase recycling of plastic-packaged water bottles in the United States. *Review of Environmental Economics and Policy, 6*(2), 190–211.

Viscusi, W. K., Huber, J., Bell, J., & Cecot, C. (2009). *Discontinous behavioural responses to recycling laws and plastic-packaged water bottle deposits.* NBER Working Paper Series, 15585.

Viscusi, W. K., Huber, J., Bell, J., & Cecot, C. (2013). Discontinuous behavioural responses to recycling laws and plastic-packaged water bottle deposits. *American Law and Economics Review, 15*(1), 110.

Walker, B. L. E. (2002). Engendering Ghana's seascape: Fanti fishtraders and marine property in colonial history. *Society & Natural Resources: An International Journal, 15*(5), 389–407.

Wall, D. (2014). *The commons in history: Culture, conflict, and ecology.* MIT Press.

Wang, Y. (2019). *Pseudo-public spaces in Chinese shopping malls: Rise, publicness and consequences.* Routledge.

Water Resources Commission. (2012). *National integrated water resources management (IWRM) Plan.*

Watson, V. (2014). African urban fantasies: dreams or nightmares? *Environment and Urbanization, 26*(1), 215–231.

Webb, D. (2017). *Critical urban theory, common property, and "the political": Desire and drive in the city.* Routledge.

West Africa Oil Watch. (2014). *Facts, West Africa oil: Some facts and figures.* http://westafricaoilwatch.org/about-us/facts/

Weszkalnys, G. (2018). Review of "Life in the time of oil: A pipeline and poverty in Chad." *Africa, 88*(3), 634–635.

What companies are for. (2019, August 24). *The Economist,* 9–10.

Widerquist, K. P., & Howard, M. W. (Eds.). (2012a). *Alaska's permanent fund dividend: Examining its suitability as a model.* Palgrave Macmillan.

Widerquist, K. P., & Howard, M. W. (Eds.). (2012b). *Exporting the Alaska model: Adapting the permanent fund dividend for reform around the world.* Palgrave Macmillan.

Wiegratz, J. (2016). *Neoliberal moral economy: Capitalism, socio-cultural change and fraud in Uganda.* Rowman and Littlefield.

Wiegratz, J. (2019). "They're all in it together": The social production of fraud in capitalist Africa. *Review of African Political Economy, 46*(161), 357–368.

Wilde, A., Adams, I., & English, B. (2013). *Fueling the future of an oil city. A tale of Sekondi-Takoradi in Ghana.* Global Communities. http://www.globalcommunities.org/publications/2013-ghana-fueling-the-future-of-an-oil-city.pdf

Wilkinson, R., & Pickett, K. (2019). *The inner level: How more equal societies reduce stress, restore sanity and improve everyone's well-being.* Penguin.

Williams, T. O., Gyampoh, B., Kizito, F., & Namara, R. (2012). Water implications of large-scale land acquisitions in Ghana. *Water Alternatives, 5*(2), 243–259.

Williamson, O. E. (1981). The modern corporation: origins, evolution, attributes. *Journal of Economic Literature, 19*(4), 1537–1567.

Williamson, O. E. (2002). The theory of the firm as governance structure: From choice to contract. *Journal of Economic Perspectives, 16*(3), 171–195.

Williamson, O. E. (2009, December 8). *Transaction cost economics: The natural progression* [Nobel Prize lecture]. The Nobel Prize. https://www.nobelprize.org/prizes/economic-sciences/2009/williamson/lecture/

Windfall tax dropped – Prez Mahama. (2014). *Ghana Trade.* http://ghanatrade.com.gh/Latest-News/windfall-tax-dropped-prez-mahama.html

Wisborg, P. (2012, April 23–26). *Justice and sustainability: Resistance and innovation in a transnational land deal in Ghana* [Paper presentation]. Annual World Bank Conference on Land and Poverty, Washington DC, United States.

Wiseman, J. (1984). Foreword. In D. R. Denman (Ed.), *Markets under the sea?* (pp. xii–xv). Institute of Economic Affairs.

World Bank. (1975). *Land reform sector policy paper.*

World Bank. (2003). *World Development Report: Making services work for poor people.* Oxford University Press.

World Bank. (2009), *Reshaping economic geography.*

World Bank. (2015). *Republic of Côte d'Ivoire.* Côte d'Ivoire Urbanization Review, Report No. AUS10013.

World Bank. (2016). *Earth observation for water resource management: Current use and future opportunities for the water sector.*

Wright, E. O. (2010). *Envisioning real utopias.* Verso.

Yang, C.-J. (2009). An impending platinum crisis and its implications for the future of the automobile. *Energy Policy, 7,* 1805–1808.

Yaro, J. A., & Tsikata, D. (2013). Savannah fires and local resistance to transnational land deals: The case of Dipale in Northern Ghana. *African Geographical Review, 32*(1), 72–87.

Yeboah, E., & Obeng-Odoom, F. (2010). "'We are not the only ones to blame": District Assemblies' perspectives on the state of planning in Ghana. *Commonwealth Journal of Local Governance, 7*(November), 78–98.

Yifeng, W. (2008). Theory of property rights: Comparing Marx and Coase. *Social Sciences in China, 29*(2), 5–17.

York, R. (2006). Ecological paradoxes: William Stanley Jevons and the paperless office. *Human Ecology Review, 13*(2), 143–147.

Zaman, A. (2019). Islam's gift: An economy of spiritual development. *American Review of Economics and Sociology, 78*(2), 443–491.

Zen, I. S., Noor, Z. Z., & Yusuf, R. O. (2014). The profiles of household solid waste recyclers and non-recyclers in Kuala Lumpur, Malaysia. *Habitat International, 42,* 83–89.

Zen, I. S., & Siwar, C. (2015). An analysis of household acceptance of curbside recycling scheme in Kuala Lumpur, Malaysia. *Habitat International, 47,* 248–255.

Zhang, L., & Bezemer, D. (2016). Finance and growth in China, 1995–2013: More liquidity or more development? *Cambridge Journal of Regions, Economy and Society, 9,* 613–631.

Zhang, L.-Y. (2015). *Managing the city economy: Challenges and strategies in developing countries,* Routledge.

Zhang, Y. (2017). *Governing the commons in China.* Routledge.

Zhang, Y. (2018). Crossing the divide: an integrated framework of the commons. *Evolutionary and Institutional Economics Review, 15*(1), 25–48.

Zimbabwe clearances condemned. (2005, June 24). SBS. http://www.sbs.com.au/news/article/2005/06/24/zimbabwe-clearances-condemned

Zouache, A. (2017a). Institutions and the colonization of Africa: some lessons from French colonial economics. *Journal of Institutional Economics,* 1–19.

Zouache, A. (2017b). Race, competition, and institutional change in J. R. Commons. *The European Journal of the History of Economic Thought, 24*(2), 341–368.

Zouache, A. (2020). From inequality to stratification: Obeng-Odoom's contribution to the study of inequality in Africa. *African Review of Economics and Finance, 12*(1), 299–306.

Index

Milton Keynes UK
Ingram Content Group UK Ltd.
UKHW020212280924
448804UK00003B/40/J